Human Genetic Diseases
Immobilised Cells and Enzymes
Iodinated Density Gradient Media
Light Microscopy in Biology
Lipid Analysis
Lipid Modification of Proteins
Liposomes
Lymphocytes
Lymphokines and Interferons
Mammalian Cell Biotechnology
Mammalian Development
Medical Bacteriology
Medical Mycology
Microcomputers in Biochemistry
Microcomputers in Biology
Microcomputers in Physiology
Mitochondria
Molecular Genetic Analysis of Populations
Molecular Neurobiology
Molecular Plant Pathology I and II
Monitoring Neuronal Activity
Mutagenicity Testing
Neural Transplantation
Neurochemistry
Neuronal Cell Lines
Nucleic Acid and Protein Sequence Analysis
Nucleic Acid Hybridisation
Nucleic Acids Sequencing
Oligonucleotide Synthesis
PCR
Peptide Hormone Action
Peptide Hormone Secretion
Photosynthesis: Energy Transduction
Plant Cell Culture
Plant Molecular Biology
Plasmids
Pollination Ecology
Postimplantation Mammalian Embryos
Preparative Centrifugation
Prostaglandins and Related Substances
Protein Architecture
Protein Engineering
Protein Function
Protein Purification Applications
Protein Purification Methods
Protein Sequencing
Protein Structure
Protein Targeting
Proteolytic Enzymes
Radioisotopes in Biology
Receptor Biochemistry
Receptor–Effector Coupling
Receptor–Ligand Interactions
Ribosomes and Protein Synthesis
Signal Transduction
Solid Phase Peptide Synthesis
Spectrophotometry and Spectrofluorimetry
Steroid Hormones

Teratocarcinomas and Embryonic Stem Cells
Transcription and Translation
Virology
Yeast

Protein Targeting
A Practical Approach

Edited by

ANTHONY I. MAGEE

National Institute for Medical Research
The Ridgeway, Mill Hill
London NW7 1AA, UK

and

THOMAS WILEMAN

Laboratory of Molecular Immunology
Dana-Farber Cancer Institute
Boston MA 02115, USA

at
OXFORD UNIVERSITY PRESS
Oxford New York Tokyo

Oxford University Press, Walton Street, Oxford OX2 6DP

Oxford is a trade mark of Oxford University Press

Published in the United States
by Oxford University Press, New York

British Library Cataloguing in Publication Data
A cataloguing record for this title is
available from the British Library

Library of Congress Cataloging in Publication Data
Protein targeting: a practical approach/edited by Anthony I. Magee
and Thomas Wileman.
(Practical approach series)
Includes bibliographical references and index.
1. Proteins—Physiological transport. 2. Proteins—Secretion.
I. Magee, Anthony I. II. Wileman, Thomas. III. Series.
QP551.P6977327 1992 574.87—dc20 91–48178
ISBN 0–19–963206–5
ISBN 0–19–963210–3 (pbk.)

Typeset by Cambrian Typesetters, Frimley, Surrey
Printed in Great Britain by
Information Press Ltd, Eynsham, Oxford

Preface

EUKARYOTIC cells are highly organized and specialized for their respective functions. This organization has required the creation of a complex spatially-restricted collection of organelles which are often bounded by lipid membranes, thus allowing the establishment of segregated intracellular environments. In order to set up and maintain these organelles the selective transport of proteins, lipids, nucleic acids, and other macromolecules has to be achieved. The mechanisms underlying these transport processes are of fundamental importance for our understanding of cellular physiology, tissue, and organ functions. Modern molecular cell-biological techniques have enabled rapid progress to be made in this area. The contributions presented here have been selected to provide detailed experimental approaches which cover many of the systems currently under study.

The first chapter deals with the entry of extracellular molecules into cells via receptor-mediated endocytosis. This is followed by two chapters describing the directed transport of molecules within polarized cells of the epithelial and neuronal type respectively. Fundamental aspects of constitutive intracellular transport can be addressed in simple unicellular eukaryotes using the powerful tools of yeast genetics, as described in the fourth chapter. The actual signals for protein targeting and the protein–protein interactions involved are often obscure and refractory to classical biochemical analysis; the fifth chapter describes a novel immunological approach to identifying such interactions. Chapters 6 and 7 describe the methodologies for studying transport of macromolecules into two of the major ubiquitous organelles, the mitochondrion and the nucleus. In order to investigate the basic biochemical nature of transport and targeting processes, reductionist approaches have been developed. Thus the next two chapters deal with *in vitro* analysis of transport events in the endocytic and secretory pathways respectively. These demonstrate many common features as well as differences between intracellular trafficking routes. The final chapter deals with lipid-derived post-translational modifications, which are emerging as major mechanisms of protein targeting.

While a book of this kind cannot cover exhaustively all aspects of protein targeting, it is hoped that the systems described will provide the background for investigators to study not only those systems themselves, but also to extrapolate the methodology to other novel areas.

London, UK — A. I. M.
Boston, USA — T. W.
December 1991

During the late stages of preparation of this book, Wayne Masterson passed away of cancer, at the early age of 31. All those who knew him as a colleague and friend will miss this outstanding young scientist and we send our deepest sympathy to his wife Claire for this sad loss.

Contents

6. Protein targeting to mitochondria 135

Ulla Wienhues, Hans Koll, Karin Becker, Bernard Guiard, and Franz-Ulrich Hartl

7. Techniques in nuclear protein transport 161

Colin Dingwall

Contributors

JOHN ARMSTRONG
Membrane Molecular Biology Laboratory, Imperial Cancer Research Fund, PO Box 123, Lincoln's Inn Fields, London WC2A 3PX, UK.

KARIN BECKER
Institut für Physiologische Chemie, Universität von München, Goethestrasse 33, 8000 München 2, Germany.

LOUISE P. CRAMER
MRC Laboratory of Molecular Cell Biology, University College, London WC1E 6BT, UK.

DANIEL F. CUTLER
MRC Laboratory of Molecular Cell Biology, University College, London WC1E 6BT, UK.

COLIN DINGWALL
Wellcome/CRC Institute, Tennis Court Road, Cambridge CB2 1QR, UK.

JULIET A. ELLIS
Department of Clinical Biochemistry, University of Cambridge, Addenbrooke's Hospital, Hills Road, Cambridge CB2 2QR, UK.

ERICA FAWELL
Membrane Molecular Biology Laboratory, Imperial Cancer Research Fund, PO Box 123, Lincoln's Inn Fields, London WC2A 3PX, UK.

JEAN-PIERRE GORVEL
European Molecular Biology Laboratory, Postfach 10.2209, Meyerhofstrasse 1, 6900 Heidelberg, Germany.

JEAN GRUENBERG
European Molecular Biology Laboratory, Postfach 10.2209, Meyerhofstrasse 1, 6900 Heidelberg, Germany.

BERNARD GUIARD
Centre de Génétique Moléculaire, Laboratoire propre du Centre National de la Recherche Scientifique, Université Pierre et Marie Curie, 91190 Gif-sur-Yvette, France.

FRANZ-ULRICH HARTL
Institut für Physiologische Chemie, Universität von München, Goethestrasse 33, 8000 München 2, Germany; present address: Laboratory of Cellular Biochemistry, Rockefeller Research Laboratory, Sloan-Kettering Institute, New York, New York 10021, USA.

MARK R. JACKMAN
Department of Clinical Biochemistry, University of Cambridge, Addenbrooke's Hospital, Hills Road, Cambridge CB2 2QR, UK.

HANS KOLL
Institut für Physiologische Chemie, Universität von München, Goethestrasse 33, 8000 München 2, Germany.

J. PAUL LUZIO
Department of Clinical Biochemistry, University of Cambridge, Addenbrooke's Hospital, Hills Road, Cambridge CB2 2QR, UK.

ANTHONY I. MAGEE
National Institute for Medical Research, The Ridgeway, Mill Hill, London NW7 1AA, UK.

WAYNE J. MASTERSON
Late of Department of Biochemistry, Medical Sciences Institute, University of Dundee, Dundee DD1 4HN, UK.

BARBARA M. MULLOCK
Department of Clinical Biochemistry, University of Cambridge, Addenbrooke's Hospital, Hills Road, Cambridge CB2 2QR, UK.

CATHERINE M. NOLAN
Department of Zoology, University College, Dublin 4, Eire.

JORGE H. PEREZ
Department of Clinical Biochemistry, University of Cambridge, Addenbrooke's Hospital, Hills Road, Cambridge CB2 2QR, UK.

ALISON PIDOUX
Membrane Molecular Biology Laboratory, Imperial Cancer Research Fund, PO Box 123, Lincolin's Inn Fields, London WC2A 3PX, UK.

RUTH SCHWANINGER
Scripps Clinic Research Foundation, Department of Molecular Biology, Mail Drop #1MM 11, 10666 Torrey Pines Road, La Jolla, CA 92037, USA.

DAVID VAUX
Sir William Dunn School of Pathology, University of Oxford, South Parks Road, Oxford OX1 3RE, UK.

ULLA WIENHUES
Institut für Physiologische Chemie, Universität von München, Goethestrasse 33, 8000 München 2, Germany.

THOMAS WILEMAN
Division of Immunology, Beth Israel Hospital, 330 Brookline Ave. RE 204, Boston, MA 02215, USA.

Abbreviations

ADH	alcohol dehydrogenase
ALP	alkaline phosphatase
ALT	alanine aminotransferase
ATP	adenosine triphosphate
BHK	baby hamster kidney (cells)
bHRP	biotinylated horseradish peroxidase
biotin-X-NHS	biotinyl-ε-aminocaproic acid *N*-hydroxysuccinimide ester
BSA	bovine serum albumin
Caco-2	human, colon-derived adenocarcinoma cell line
carbachol	carbamylcholine
CCCP	carbonyl cyanide *m*-chlorophenylhydrazone
cDNA	complementary DNA
CDR	complementarity determining region
CHO	Chinese hamster ovary (cells)
CI-Man 6-P	cation-independent mannose 6-phosphate (receptor)
CMFM	cysteine- and methionine-free DMEM
CP	creatine phosphate
CPK	creatine phosphokinase
CRD	cross-reacting determinant
DAPI	4,6′-diamidino-2-phenylindole
DCG	dense core granules
DHFR	dihydrofolate reductase
DMEM	Dulbecco's modified Eagle's medium
DMSO	dimethylsulphoxide
DTT	dithiothreitol
EDTA	ethylenediaminetetraacetate
EGF	epidermal growth factor
EGTA	ethyleneglycol-bis-(2-aminoethyl ether)*N*,*N*,*N*′,*N*′-tetraacetic acid
E^{hu}AChE	human erythrocyte acetylcholinesterase
ELISA	enzyme linked immunosorbent assay
endo D	endoglycosidase D
ER	endoplasmic reticulum
FACS	fluorescence activated cell sorter
FCA	Freund's complete adjuvant
FCS	fetal calf serum
FIA	Freund's incomplete adjuvant
FM	vesicle fusion medium
G418	'geneticin'
GIP	general insertion protein
GlcNAc	*N*-acetylglucosamine
GlcNAc Tr 1	*N*-acetylglucosamine transferase 1
G-MEM	Glasgow minimal essential medium
GPI	glycosyl phosphatidylinositol

GPI-PLC	GPI-specific phospholipase C
GPIsp	GPI signal peptide
HB	handling buffer
HIV	human immunodeficiency virus
HMG-CoA	hydroxymethylglutaryl-coenzyme A
HPLC	high performance liquid chromatography
HRP	horseradish peroxidase
HSA	human serum albumin
hsp	heat shock protein
HTC	rat hepatoma cells
IEF	isoelectric focusing
Ig	immunoglobulin
IGF-II	insulin-like growth factor II
IM	internalization medium
IPRL	isolated rat liver preparations
IPTG	isopropyl β D-thiogalactoside
K_a	affinity constant
K_d	dissociation constant
KOAc	potassium acetate
LDL	low-density lipoprotein
LTR	long terminal repeat
Man I	α-1,2-mannosidase I
Man 6-P	mannose 6-phosphate
MBS	*m*-maleimidobenzoyl-*N*-hydroxysuccinimide ester
MDCK	Madin–Darby canine kidney (cells)
MEM	minimum essential medium
MgOAc	magnesium acetate
MOPS	3-[n-Morpholino] propanesulfonic acid
M_r	relative molecular mass
MTX	methotrexate
NGF	nerve growth factor
NLM	normal lymphocyte medium
NP	nuclear pellet
NP40	Nonidet P40
NRK	normal rat kidney (cells)
OD	optical density
ORF	open reading frame
PAGE	polyacrylamide gel electrophoresis
PBS	phosphate-buffered saline
PCR	polymerase chain reaction
PEG	polyethylene glycol
p.f.u.	plaque-forming units
PI–PLC	phosphatidylinositol-specific phospholipase C
PMSF	phenylmethylsulphonyl fluoride
PNS	post-nuclear supernatant
PPO	2,5-diphenyloxazole
PVP	polyvinyl pyrollidone
R18	octadecyl-rhodamine B-chloride

RME	receptor-mediated endocytosis
RPMI	Roswell Park Memorial Institute (1640 medium)
RSV	Rous sarcoma virus
SAM	S-adenosylmethionine
SDS	sodium dodecyl sulphate
SG11	secretogranin 11
SLV	synaptic-like vesicles
SV40	simian virus 40
TBS	Tris-buffered saline
TCA	trichloroacetic acid
TCM	thymus-conditioned medium
TEA	triethanolamine
TEMED	*N*,*N*,*N*′,*N*′-tetramethyl-ethylenediamine
TES	*N*-tris(hydroxymethyl)methyl-2-aminoethane sulphonic acid
TGM	thymocyte growth medium
TGN	*trans*-Golgi network
TLC	thin layer chromatography
TMPD	*N*,*N*,*N*′,*N*′-tetramethylphenylenediamine
TPB	tryptose phosphate broth
TX100	Triton X-100
TX-114	Triton X-114
UDP	uridine diphosphate
VSG	variant surface glycoprotein
VSV	vesicular stomatitis virus
WGA	wheat germ agglutinin
5-FOA	5-fluoro-orotic acid
5′NT	5′nucleotidase

1

Receptor-mediated endocytosis and lysosomal transport

CATHERINE M. NOLAN

1. Introduction

Endocytosis is the process by which cells internalize their plasma membrane together with molecules adsorbed to the plasma membrane, fluid from the extracellular medium, and solutes dissolved in this fluid. Most cell types perform this process to greater or lesser extents. In addition to being a means of maintaining the correct proportion and composition of the plasma membrane, endocytosis can serve many other functions, including nutrition, regulation of the extracellular environment, antigen processing and presentation, and hormone processing. It may play an important role in processes such as growth and differentiation, wound healing, embryonic development, and pathological processes such as inflammation. The endocytic process can also be subverted by bacteria and viruses for penetration of them or their toxins into the host cell.

Substances dissolved in the extracellular fluid of cells are internalized by fluid-phase endocytosis while substances adsorbed to the plasma membrane are internalized by adsorptive endocytosis. A special case of adsorptive endocytosis is when molecules are specifically bound by receptors present on the cell surface and enter the cells by virtue of this interaction. The latter process is termed receptor-mediated endocytosis. It is this process and the fate of such internalized ligands and receptors that form the basis of this chapter. Receptor-mediated endocytosis has been the subject of recent reviews (1, 2). The main differences between fluid-phase and receptor-mediated endocytosis are listed in *Table 1*.

Ligand–receptor complexes form at the cell surface and are internalized from specialized areas of the plasma membrane. These are called 'coated pits' because of the presence on their cytoplasmic face of a lattice of clathrin. Coated pits pinch off to form coated vesicles which subsequently lose their clathrin coat. The ligand–receptor complexes then enter an endosomal compartment and can subsequently follow different fates. In some cases, both ligand and receptor can return to the plasma membrane from the endosome

Table 1. Comparison of ligand uptake by fluid-phase endocytosis and by receptor-mediated endocytosis

Receptor-mediated endocytosis	Fluid-phase endocytosis
Ligand binding occurs at 4 °C	No ligand binding at 4 °C
No internalization at 4 °C	No internalization at 4 °C
Receptor required	No receptor necessary
Saturable	Non-saturable
Efficient	Inefficient
Selective	Non-selective

(recycling). In other cases, ligand and/or receptor can be transported to the lysosomal compartment where they are degraded by the acid hydrolases localized there. In yet other cases, the ligand is degraded in lysosomes while the receptor is free to recycle. In polarized cells such as epithelial cells the ligand can be transported from one surface of the cell to the other. The ligand is said to be transcytosed while the receptor may or may not be degraded. All of these transport pathways have been recently reviewed (2).

As ligands are transported from the plasma membrane to lysosomes they are exposed to an environment which is increasingly acidic. The pH of the lumen of the endosomal compartment has been estimated to be 5.0–6.0 while that of the lysosomal compartment has been determined to be approximately 4.5–5.0. The low pH of these compartments is maintained by means of ATP-dependent proton pumps and is an important component of the regulation of the transport pathway (3).

2. Experimental analysis of receptor-mediated endocytosis (RME)

RME has two components which can be experimentally measured. Binding of the ligand at the cell surface is followed by internalization of the ligand into the interior of the cell.

2.1 Cell-surface ligand binding

Binding of ligands at the cell surface is a non-energy-requiring process and can be determined by measuring cell-surface associated ligand following incubation of cells with ligand at 4 °C. Molecules internalized by fluid-phase endocytosis should not show any association with the cell surface at this temperature.

Since cell surface receptors constitute only a small proportion of the total cell surface molecules, measurement of cell-surface ligand binding requires the availability of a sensitive assay. If a ligand has enzymatic activity, this may present a convenient method of measuring binding activity. The binding of lysosomal enzymes to the cation-independent mannose 6-phosphate receptor (CI-Man 6-P receptor) can be measured in this way (*Table 2*) (4). A widely used method of assaying cell surface ligand binding is to use a radiolabelled ligand. In practice, radioiodinated ligands have proved very convenient and sensitive.

A second practical requirement is a means of quickly and completely separating cell-surface-receptor bound ligand from unbound ligand. In the case of adherent cells, this can be accomplished by aspirating the binding medium from the cells and subsequently washing the cells (see *Protocol 1*). For binding studies on cells grown in suspension, the cells can be centrifuged through oil to remove unbound ligand at the end of the incubation. In this procedure, the binding medium is layered over a mixture of 4 parts silicon oil, 1 part mineral oil in a microcentrifuge tube and spun for 30 sec–4 min in a microcentrifuge (5). The tips of the tubes containing the cells are then cut off and counted.

Protocol 1 is a procedure used to measure binding of lysosomal enzymes to cells such as human fibroblasts and mouse L cells which express the CI-Man 6-P receptor (4). The general principles apply in many situations, however, and the protocol can be adapted for other ligands and cells. Essentially the same protocol was used for the experiments using ^{125}I-labelled epidermal growth factor ([^{125}I]EGF) described in this chapter.

Protocol 1. Cell surface binding of ß-glucuronidase by CI-Man 6-P receptor-expressing cells

1. Culture the cells on 35-mm Petri dishes until confluent.
2. Transfer the Petri dishes on to ice in a suitable container.
3. Aspirate the medium and wash the cells once with 2 ml of ice-cold Dulbecco's modified Eagle's medium (DMEM).
4. Wash the cells with 2.0 ml of cold (DMEM) containing 5 mg/ml human serum albumin (DMEM/HSA).
5. Add 1.0 ml of DMEM/HSA containing 20 mM Hepes and 4000 units/ml β-glucuronidase $\pm$ 5 mM mannose 6-phosphate (Man 6-P) to replicate dishes of cells.
6. Incubate for 2 h on ice.
7. At the end of the incubation, wash all dishes five times with 2.0 ml of cold phosphate-buffered saline (PBS) containing 5 mg/ml HSA (PBS/HSA).
8. Solubilize the cells in 1.0 ml of 1.0% sodium deoxycholate for 30 min at 4 °C and assay for β-glucuronidase activity (6) (see *Protocol 6*).
9. Measure the protein content of cell extracts by the Lowry assay (7) or using the BCA reagent (Pierce Biochemicals).

2.1.1 Specificity of cell-surface ligand binding

A feature of receptor-mediated binding is its specificity: ligand binding is inhibited by the presence of specific inhibitors or closely related species but is not affected by the presence of non-related molecules. In practice the specificity of ligand binding to a cell surface is determined by incubating cells with the ligand at 4 °C in the presence and absence of the putative specific inhibitor. 'Specific binding' is defined as the total ligand bound in the absence of inhibitor (total binding) minus the amount of ligand bound in the presence of an excess of inhibitor (non-specific binding). Ligand bound in the presence of the specific inhibitor represents ligand associated with the cell surface by a non-receptor-mediated interaction or sticking to the experimental container, etc. Experimental conditions should be chosen which minimize this non-specific binding.

Table 2 shows the specific binding of ß-glucuronidase to the human CI-Man 6-P receptor expressed in mouse L cells (4). Binding of lysosomal enzymes to this receptor is mediated through man 6-P residues present on the oligosaccharide side chains of the enzyme and is inhibited by the presence in the binding medium of the competing sugar Man 6-P. In the case of the experiment shown in *Table 2*, specific (Man 6-P-dependent) binding of ß-glucuronidase is the total binding (491 units/mg) less the non-specific binding (164 units/mg), that is 327 units/mg. Note also that, in cells that do not express the CI-man 6-P receptor, there is no Man 6-P-dependent binding of the ligand.

Table 2. Binding of ß-glucuronidase to CI-Man 6-P receptor-negative mouse L cells and to L cells expressing the recombinant human receptor[a]

	ß-glucuronidase activity (units/mg protein)	
Addition to binding medium	**Receptor-negative L cells**	**Receptor-positive L cells**
None	23	63
Human ß-glucuronidase (4000 units)	28	491
Human ß-glucuronidase (4000 units) + 5 mM Man 6-P	29	164
Man 6-P inhibitable ß-glucuronidase binding	—	327

[a] Reproduced with permission from reference 4.

When the determinants of specificity are unknown, a convenient way to estimate specific binding is to radiolabel the ligand of interest and demonstrate that at 4 °C, binding of this ligand can be inhibited by the presence of an excess of the non-radiolabelled ligand but not by the presence of unrelated ligands. The binding of [^{125}I]EGF to Hep 2 cells is shown in *Table 3*. This binding is inhibited by the presence of excess unlabelled EGF but not by the unrelated ligands, insulin and transferrin.

Table 3. Specificity of binding of [^{125}I]EGF to Hep 2 cells

Competing ligand	**Radioactivity/dish**[a] **(c.p.m. × 10^{-2} ± SEM)**
None	1939 ± 131
EGF	85 ± 135
Transferrin	1899 ± 58
Insulin	1899 ± 185

[a] Hep 2 cells were incubated for 2 h at 4 °C with [^{125}I]EGF (0.5 ng/ml) in the absence of any competing ligand or in the presence of 100 ng/ml unlabelled EGF, 1 μg/ml transferrin, or 1 μg/ml insulin. Values presented are averages of three dishes with the standard error of the mean (SEM).

2.1.2 Receptor-mediated cell surface binding is saturable

Because of the presence of a finite number of receptors at the plasma membrane, cell surface ligand binding shows a marked dependence on ligand concentration. At low ligand concentrations, a linear increase in binding is seen; at higher ligand concentrations saturation occurs. *Figure 1* shows the saturability of β-glucuronidase binding to cells expressing the CI-Man 6-P receptor.

2.1.3 Cell surface binding can be measured at 37 °C

If it is necessary to look at cell surface binding at temperatures higher that 4 °C, this can be achieved by performing the binding under conditions where internalization is inhibited by using inhibitors of energy production (8). Cells are pre-incubated with PBS containing 1 mg/ml HSA and 10 mM sodium azide, 10 mM sodium fluoride, and 1 mM sodium cyanide, for 10–30 min before ligand binding. Binding is subsequently performed in binding medium containing these inhibitors.

2.1.4 Determination of receptor numbers and binding constants

Estimates of receptor numbers per cell and of binding characteristics, such as affinity and dissociation constants (K_a and K_d), can be obtained from an experiment such as that shown in *Figure 1*. The data are replotted as B (on the x-axis) against B/F (on the y-axis) where B (bound) is the concentration of ligand in the incubation present as ligand–receptor complex at equilibrium and F (free) is the concentration of free ligand present in the incubation at equilibrium. For a simple bimolecular reaction, a straight line is obtained where the slope $= -1/K_d$ or $-K_a$. The x-intercept is an estimate of B_{max}.

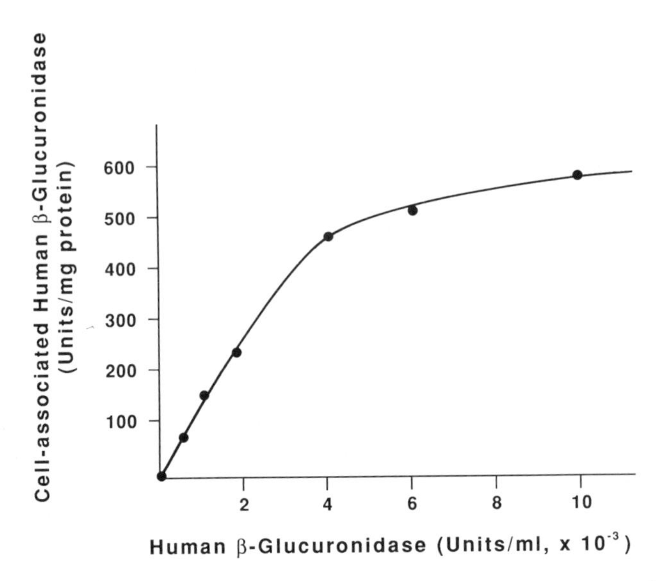

Figure 1. Binding of human ß-glucuronidase to CI-Man 6-P receptor-expressing mouse L cells. Mouse L cells expressing the recominant human CI-Man 6-receptor (4) were incubated at 4 °C with increasing concentrations of human ß-glucuronidase, in the absence and presence of 5 mM Man 6-P. For each enzyme concentration, specific (Man 6-P-dependent) binding of ß-glucuronidase was determined and plotted.

Such an analysis is generally referred to as a Scatchard analysis (9). For a good comprehensive treatment of the analysis of binding data, see reference 10.

2.2 Ligand internalization

When cells are exposed to ligand at 37 °C for a given time, the ligand associated with the cells comprises ligand which has been internalized into an intracellular location(s) in addition to ligand associated with the plasma membrane receptor. The relative proportions of these two classes of ligand (cell-surface and internalized) vary depending on the experimental conditions chosen, as well as on the nature of the receptor–ligand interaction. It is often useful to be able to quantitatively distinguish cell surface from internalized ligand.

Various methods have been described which enable this distinction to be made. These usually involve the removal of cell-surface bound ligand by a method which does not lyse the cells, thus leaving the internalized ligand cell-associated. In these methods, cells are first incubated at 37 °C, cooled to prevent any further internalization of the ligand, and washed to remove non-receptor bound ligand. Cell-surface bound ligand can often be removed by changing some parameter of the binding medium. For example, if ligand binding requires divalent cations, incubating cells with medium containing 20 mM EGTA may reverse it (11). Incubating cells which have bound lysosomal enzymes through the CI-Man 6-P receptor with 5 mM Man 6-P for 15 min at 4 °C will remove cell-surface receptor bound enzyme (12). Similarly, dextran sulphate (4 mg/ml) or heparin (10 mg/ml) have been used to remove low-density lipoprotein (LDL) bound to the LDL-receptor (13).

Two methods of removing cell-surface bound ligand that have relatively broad applicability are the so-called 'acid-stripping' technique and the proteolytic treatment of cell surfaces.

2.2.1 Acid stripping of cell-surface bound ligand

This technique depends on the observation that the binding of many ligands which are internalized by receptor-mediated endocytosis shows a marked pH dependence: ligand binds to receptor at physiological pH but bound ligand dissociates at acidic pH. Many reports of acid-stripping techniques involve the use of relatively harsh conditions, for example, incubation of cells with 0.2 M acetic acid, pH 2.5, containing 0.5 M NaCl, for 6 min (14). While these conditions effectively remove cell-surface bound ligand, the cells are often damaged by the procedure and are not of use in following further processing of the ligand.

A gentler protocol described by Ascoli (15), using a glycine buffer, has been shown to be very useful in selectively removing cell-surface bound ligand (*Protocol 2*). In addition, this procedure has been shown to have very little effect on the rebinding of ligand and on the ability of cells to subsequently process the ligand normally. Using this protocol more than 90% of [^{125}I]EGF bound to human fibroblasts at 4 °C can be removed. When the incubation is at 37 °C, however, only about 20% of ligand can be removed, the remainder (internalized) being resistant to the acid treatment.

Protocol 2. Removal of cell-surface bound ligand by acid-stripping

1. Incubate the cells (in 35-mm Petri dishes) at 37 °C with ligand (e.g. [^{125}I]EGF).
2. Cool the cells on ice and wash them four times with 2.0 ml of PBS containing 5 mg/ml bovine serum albumin (PBS/BSA).
3. Incubate the cells with 1.0 ml of ice-cold 50 mM glycine, 100 mM NaCl, pH 3.0 (acid wash buffer) for 2 min.
4. Remove the solution and retain it.[a]
5. Wash the dishes with a further 1.0 ml of the acid wash buffer and combine this wash with the first.
6. Solubilize the cells using 1.0 ml of 1 N NaOH (in the case of ^{125}I-labelled ligands) or appropriate detergent solution.

Protocol 2. *Continued*

7. Determine the amount of ligand in the pooled acid washes and in the cell extracts.

[a] If an enzyme activity is being measured rather than radioactivity, check that the enzyme activity is stable to the acid treatment and/or neutralize the acid washings immediately on removal from the cells.

2.2.2 Removal of cell-surface bound ligands by proteolytic treatment of cells

The use of proteases to remove cell-surface bound ligands has the advantage over acid-stripping that it does not depend on the binding characteristics of the ligand–receptor system. As long as the ligand is susceptible to protease it should be possible to remove it by limited proteolysis. It does have two disadvantages, however.

- The receptor may also be degraded and so the ability to re-bind ligand can be significantly reduced.
- Care must be taken that adherent cells are not detached from the dish by the proteolytic enzyme, if subsequent incubations are necessary.

Several methods of removing cell-surface bound ligand from cells by proteolysis using trypsin, proteinase K, or a mixture of chymotrypsin and proteinase K have been described (5, 16, 17). In general, the incubation of cells with trypsin (0.1% w/v) (Sigma type III) in Hank's balanced salt solution for 10–30 min at 4 °C gives good results (5). To stop the reaction, use soybean trypsin inhibitor (Sigma type I-S) at 0.1–0.2% (w/v).

3. Degradation of ligand

The fate of many ligands following internalization is that they are transported to lysosomes and are degraded. *Figure 2* shows the association of [^{125}I]EGF with human fibroblasts during a prolonged incubation of the cells with the ligand at 37 °C. The cell-associated radioactivity increases to a maximum but then declines as the ligand is degraded.

Ligand degradation can be conveniently monitored if the ligand is radioiodinated, since the products of such degradation are iodoamino acids. These accumulate in the extracellular medium and can be quantitated by precipitation of the medium with trichloroacetic acid (TCA). In concentrations of TCA greater than 10%, high molecular weight proteins and peptides are quantitatively precipitated while low molecular weight species such as iodoamino acids remain soluble.

Protocol 3. TCA precipitation of ^{125}I-labelled species

Materials

- 5 mg/ml BSA
- 100% (w/v) TCA
- 20% (w/v) TCA

Procedure

1. Place the sample containing ^{125}I-labelled species in a tube which can be counted in the available γ-counter. The volume of the sample depends on the c.p.m. present.
2. Count the tubes.[a] This is the 'total radioactivity'.
3. Add an aliquot of 5 mg/ml BSA as carrier protein.[b] The protein concentration should be at least 10 μg/ml.
4. Add one-quarter volume of ice-cold 100% (w/v) TCA such that the final acid concentration is 20%. You should see a strong precipitate.
5. Incubate for at least 1 h (can be left longer or overnight) on ice.
6. Centrifuge at 4 °C to form a pellet. In a microcentrifuge, 2 min at top speed should be sufficient. In a Sorvall RT6000B, use a 1000–3000 r.p.m. spin for 15–20 min.
7. Aspirate the supernatant taking care to retain the pellet.
8. Wash the pellet with 1.0 ml of ice-cold 20% TCA. Centrifuge and aspirate the supernatant.
9. Count the pellet. This is the 'TCA-precipitable radioactivity'. TCA-soluble radioactivity is the 'total radioactivity' less the 'TCA-precipitable radioactivity'.

[a] Alternatively, the precipitations can be set up in duplicate and replicates can be counted at the end, without aspiration of the first supernatant, to determine the 'total radioactivity'.
[b] This step may not be necessary if the ^{125}I-labelled sample has a high protein concentration but carrier protein should always be added when protein concentration is low.

Figure 3 shows the degradation of [^{125}I]EGF by human fibroblasts as measured by the disappearance of radioactivity from cells and its appearance as TCA-soluble radioactivity in the medium. To confirm that the ligand is in fact being degraded in a lysosomal compartment, one should demonstrate that the degradation is inhibited by one or more of the known inhibitors of lysosomal degradation (see Section 7). In *Figure 3* inhibition by the amine, ammonium chloride, is shown (see Section 7.2).

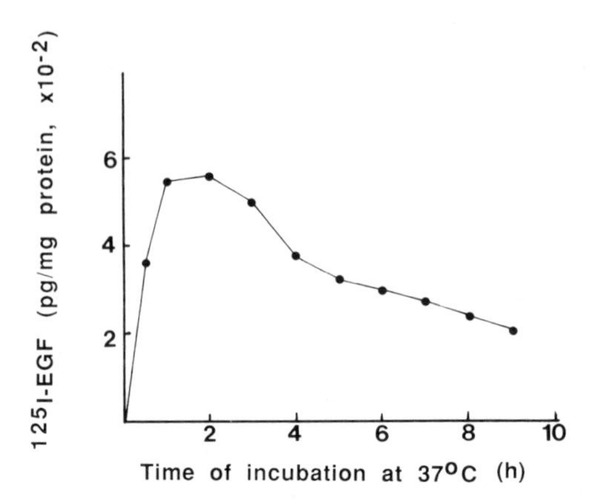

Figure 2. Time course of association of [^{125}I]EGF with human fibroblasts. Human fibroblasts were incubated at 37 °C with [^{125}I]EGF (10 ng/ml) for the indicated times. The medium was then aspirated; the cells were washed with PBS/HSA and solubilized in 1M NaOH. Cell-associated radioactivity is represented as pg [^{125}I]EGF/mg cell protein.

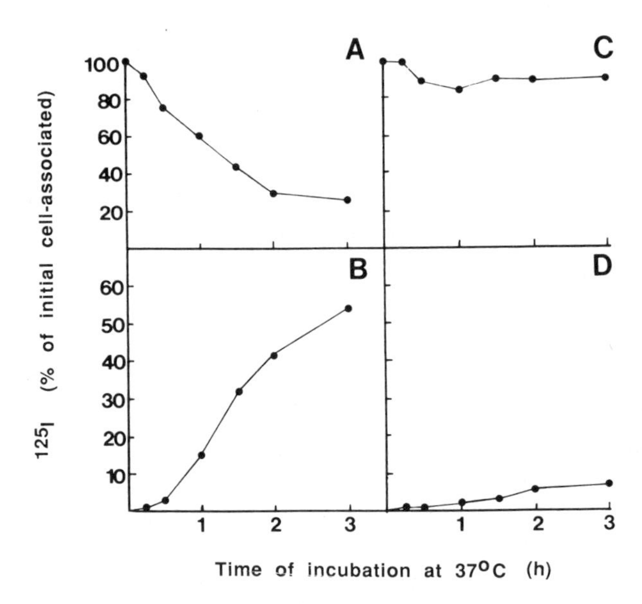

Figure 3. Degradation of [^{125}I]EGF by human fibroblasts and the inhibitory effect of NH_4Cl. Cells were pre-incubated for 1 h at 37 °C with [^{125}I]EGF, in the presence of 10 mM NH_4Cl (to prevent any degradation during the uptake). They were then washed and incubated in fresh medium in the absence (A, B) or presence (C, D) of 10 mM NH_4Cl. At the indicated times, medium was removed and analysed for TCA-soluble radioactivity (B, D) and the cells were solubilized and counted for radioactivity (A, C). A small amount of TCA-precipitable radioactivity (max. 10%) appeared in the medium during the incubation and is not shown in this figure. This probably represents the small amount of this ligand which is thought to be recycled intact to the medium (see Section 4).

3.1 Ligand degradation can occur in endosomes

Recently, it has been shown that acid hydrolases may be present in endosomes in addition to their location in lysosomes (18). This can result in the appearance of degradation products in the medium of cells prior to the transfer of ligand to lysosomes. Such endosomal degradation occurs relatively fast and is sensitive to the effects of inhibitors of pH gradients, such as weak base amines and ionophores (see Section 7.2), and also to specific inhibitors of lysosomal enzymes (see Section 7.3).

3.2 Degradation of ligand and recycling of receptor

Early in the study of receptor-mediated endocytosis, it was observed that in some cases (for example, the Man 6-P receptor and the mannose receptor) the total number of ligands internalized and degraded exceeded the total number of receptors present at the plasma membrane and in intracellular compartments (5, 8). This was true even in the presence of the protein synthesis inhibitor cycloheximide and it was shown that each receptor was capable of being used more than once. The receptors are said to recycle. Other examples of receptors which recycle while their ligand is degraded are the asialoglycoprotein receptor (19) and the LDL-receptor (20).

Receptors which recycle usually exist as a pool of receptors in an intracellular compartment in addition to those present at the plasma membrane (21, 22). The relative sizes of these compartments varies with the receptor and with the cell type. The existence of an intracellular pool of receptors can be demonstrated in three ways:

- trypsin treatment of cells at 4 °C (see Section 2.2.2) followed by warming of the cells to 37 °C to allow the movement of receptors to the cell surface from an intracellcular compartment (5);
- comparing binding of ligand to the cell surface with binding to a total membrane preparation derived from the cells (21);
- performing ligand-binding studies on cells which have been permeabilized with a detergent such as saponin (4, 22).

In cells expressing the CI-Man 6-P receptor, the intracellular pool of the receptor comprises 80–90% of the total cellular receptor and becomes accessible to ligand at 4 °C when the cells are treated with saponin. *Figure 4* shows the binding of [^{125}I]insulin-like growth factor II (another ligand of this receptor; reference 23) to the CI-Man 6-P receptor in permeabilized and non-permeabilized mouse L cells (4).

4. Recycling of ligand and receptor

The recycling of transferrin has been observed by incubating cells with radioiodinated differic transferrin for 1 h at 37 °C, washing to remove

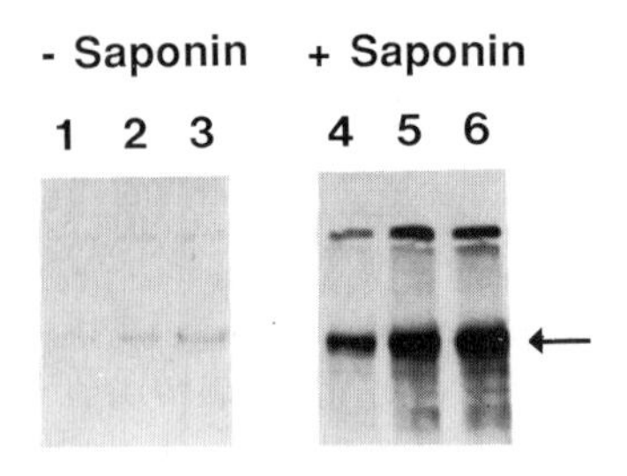

Figure 4. Cells expressing the CI-Man 6-P receptor have both a plasma membrane and an intracellular pool of receptor. Mouse L cells expressing the recombinant human CI-Man 6-P receptor/IGF-II receptor were pre-incubated for 30 min in the absence (lanes 1–3) or presence (lanes 4–6) of 0.25% (w/v) saponin. The cells were incubated at 4 °C with [^{125}I]IGF-II at 25 ng/ml (lanes 1, 4), 50 ng/ml (lanes 2, 5), and 100 ng/ml (lanes 3, 6). They were then washed and exposed to the cross-linking reagent, disuccinimidyl suberate. Following quenching of the reaction, the cells were analysed by SDS–polyacrylamide gel electrophoresis and fluorography (4). The arrow indicates the position of the receptor.

unbound ligand, and then continuing the incubation in fresh medium containing excess unlabelled transferrin (24). Under these conditions, the majority of internalized radioactivity appeared in the extracellular medium. Unlike the situation seen with ^{125}I-EGF (*Figure 3*) where the majority of radioactivity recovered in the medium was TCA-soluble, the radioactivity recovered in the medium in the ^{125}I-transferrin experiment consisted of TCA-precipitable material (24).

It has also been shown that a small percentage of a cohort of internalized ligands that are normally degraded (such as EGF, LDL, asialoorosmucoid, etc.) can be recycled intact to the medium. This has been variously called diacytosis and retroendocytosis (1).

5. Transcytosis

In polarized cells, such as epithelial cells, ligands and receptors can be transported from one region of the plasma membrane to another. The best characterized example of this is the transcytosis of polymeric IgA from the basolateral surface of Madin–Darby canine kidney (MDCK) cells to the apical surface (25).

To study transcytosis experimentally, cells have been grown on permeable filters which allow manipulation of the cells from both the apical and basolateral side. Usually ligand is added to the medium on one side of the filter and the opposing medium is then examined for the presence of the ligand. As polarized cells are discussed in detail in Chapter 2 of this volume, I will not discuss transcytosis any further here.

6. Subcellular fractionation of cells on density gradients

In disrupted cells, the components of the RME/lysosomal transport system have different densities, with vesicles derived from the lysosomal compartment being denser than those derived from the endosomal compartment or from the plasma membrane. Subfractionation of disrupted cells on density gradients has allowed the movement of ligands along these transport pathways to be monitored. A convenient method of performing such fractionations is by means of the self-forming gradient medium, Percoll (Pharmacia) (26, 27).

Protocol 4. Subcellular fractionation of cells on Percoll density gradients

Materials

- homogenizing buffer: 0.25 M sucrose, pH 7.3 with 1 M KOH
- Percoll stock: dilute 49.5 ml Percoll with 5.5 ml of 2.5 M sucrose and adjust the pH to 7.3 using 1 M HCl (requires approx 200 μl 1 M HCl—mix gently, do not use a magnetic stirring bar)
- working Percoll solution: dilute 3 ml of stock Percoll with 7 ml of homogenizing buffer. This results in a solution with a density of approximately 1.07 g/ml.

Method

All steps are performed on ice or at 4 °C.

1. Use a 10-cm dish of 80–90% confluent human fibroblasts.
2. Wash the cell monolayers twice, using 5.0 ml of homogenizing buffer.
3. Gently scrape the monolayer off the dish in 5 ml of homogenizing buffer using a rubber policeman. Centrifuge at 800 *g* for 10 min.
4. Resuspend the cell pellet in 1 ml of homogenizing buffer and homogenize by 6–10 strokes in a loose-fitting Dounce homogenizer.
5. Centrifuge at 800 *g* for 10 min. Retain the supernatant.
6. Wash the pellet with 1 ml of homogenizing buffer. Centrifuge at 800 *g* for 10 min and combine this supernatant with the first.
7. Place 9 ml of the working Percoll solution in a suitable centrifuge tube (e.g. Beckman ultra-clear, 16 × 76 mm). Carefully layer the combined supernatants on top.

Protocol 4. *Continued*

8. Centrifuge at 33 000 *g* in a 50Ti rotor for 50 min.

9. Collect 10-drop (approximately 0.4 ml) fractions by puncturing the bottom of the tube with a needle attached to a length of capillary tubing.

6.1 Characterization of gradient fractions and markers of subcellular fractions

6.1.1 Density

The density of the individual fractions can be determined by accurately weighing aliquots of the fractions. Alternatively, marker beads of known density are supplied by Pharmacia and can be used to characterize the density of a gradient run in parallel.

6.1.2 Plasma membrane

The location of the plasma membrane can be determined by measuring the distribution of an enzyme such as 5′-nucleotidase which is characteristic of the plasma membrane (28). The use of this enzyme to locate the position of plasma membrane is illustrated in *Figure 5*. In practice, a portion of such enzymes may be in an intracellular location due to the normal recycling of the plasma membrane. Another method which can be used to monitor the distribution of the plasma membrane in subcellular fractions is to radioiodinate the cell surface at 4 °C with lactoperoxidase prior to fractionation (26). The subcellular fractions are then simply counted for radioactivity.

Protocol 5. Assay of 5′-nucleotidase

This assay involves measuring the release of inorganic phosphate from AMP. The protocol given is a modification of that of Aaronson and Touster (28).

Reagents

- reagent A: Na-AMP, 50 mM, pH 7.0
- reagent B: Glycine–NaOH buffer, 0.5 M, pH 9.1
- reagent C: $MgCl_2$, 0.1 M
- TCA 8% (w/v)
- 2.5% (w/v) ammonium molybdate in 2.5 M H_2SO_4
- Fiske–Subbarow reagent (Fisher Chemical Co.)

Prepare the assay reagent by combining solutions (A), (B), (C), and distilled water in a volume ratio of 1:2:1:5. The resulting reagent can be stored frozen for at least a year.

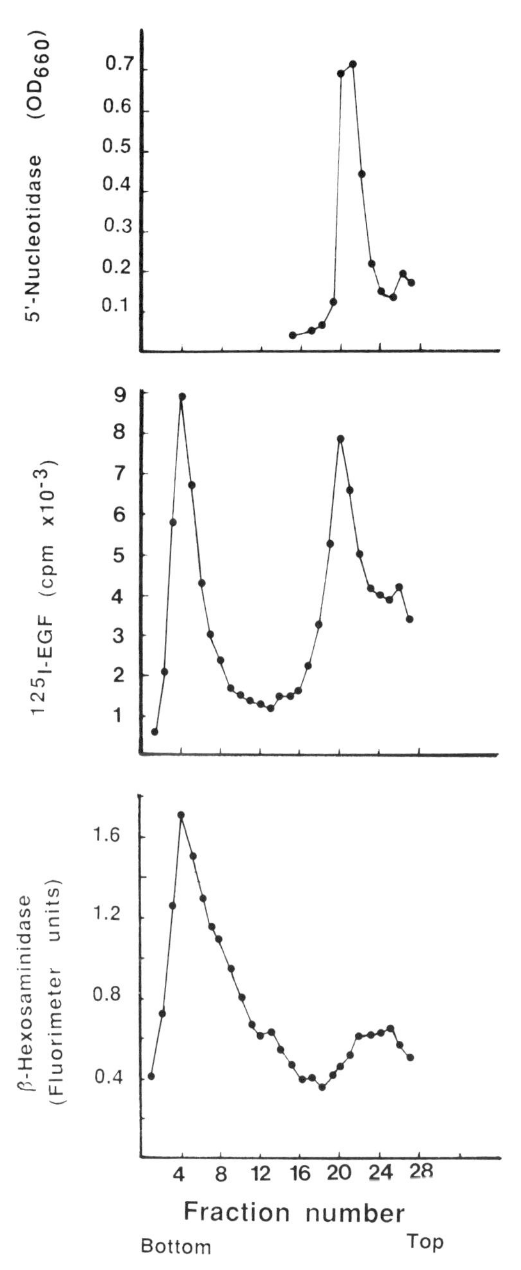

Figure 5. Percoll density gradient fractionation of human fibroblasts. Human fibroblasts on a 10-cm Petri dish were incubated with [^{125}I]EGF for 60 min at 37 °C. They were then washed, homogenized, and subjected to subcellular fractionation on a Percoll density gradient, as described in *Protocol 4*. The fractions were counted for radioactivity; aliquots were assayed for the plasma membrane marker 5′-nucleotidase and the lysosomal marker ß-hexosaminidase. [^{125}I]EGF serves as a marker of both endosomal and lysosomal positions. (In this gradient, the plasma membrane and endosomal components are not widely separated.)

Protocol 5. *Continued*

Method

1. Add an aliquot (50–100 μl) of the gradient fractions to 0.45 ml of the assay reagent in 12 × 75 mm borosilicate tubes.
2. Incubate at 37 °C for up to 2 h. Include a suitable control tube (containing water and assay reagent) and a phosphate standard (0.1–0.5 μM of sodium phosphate).
3. Stop the assay by adding 2.5 ml of ice-cold 8% TCA. Centrifuge at 3000 *g* at 4 °C for 5 min.
4. Transfer 2.0-ml aliquots of the supernatant to fresh tubes.
5. Add 0.5 ml of 2.5% solution of ammonium molybdate in 2.5 M H_2SO_4 and mix.
6. Add 0.2 ml of Fiske–Subbarow reagent and mix.
7. After 10 min measure the optical density at 660 nm (OD_{660}) against the water blank.

One unit of enzyme activity is defined as that amount of enzyme that catalyses the release of 1 μmol of inorganic phosphate per min. Alternatively, the activity can be expressed as OD units.

6.1.3 Endosomes

As yet, there are no enzyme markers which can be described as uniquely characteristic of endosomes. As mentioned in Section 3.1, in some cells enzymes originally characterized as lysosomal enzymes have also been shown to occur in endosomes (18). The position of endosomes in subcellular fractions may be determined by incubating the cells with a convenient ligand prior to the fractionation (26). If a suitable time of incubation is chosen, the ligand can be localized to endosomes and/or lysosomes. A useful ligand for both endosomal and lysosomal determination is [^{125}I]EGF (illustrated in *Figure 5*) while [^{125}I]transferrin is useful for endosome localization.

6.1.4 Lysosomes

Lysosomes contain many hydrolytic enzymes whose activity can be conveniently measured. These include *β*-glucuronidase, acid phosphatase, *β*-galactosidase, and α-mannosidase (26). Of these, *β*-hexosaminidase is often the enzyme of choice as a marker enzyme since it is relatively abundant.

Protocol 6. Assay of *ß*-hexosaminidase[a]

This is a sensitive fluorometric assay (29) and requires a fluorimeter equipped with appropriate filters (excitation at 360 nm and emission at 450 nm).

Materials

- substrate: 4-methylumbelliferyl-2-acetamido-2-deoxy-β-D-glucopyranoside at 5 mM in 0.02 M sodium phosphate-citrate, pH 4.4; 0.1% (w/v) saponin. This can be stored frozen for several months. The substrate tends to precipitate out of solution on storage. On thawing, warm gently to dissolve the substrate completely.
- reaction stopping buffer: 0.32 M glycine; 0.2 M sodium carbonate, pH 10.5

Method

1. Place 25-μl aliquots of the subcellular fractions in suitable tubes.
2. Add 100 μl of the substrate reagent. Incubate at 37 °C for 15–30 min.
3. Place the tubes on ice and add 1.85 ml of glycine/carbonate stopping buffer. Vortex the tubes.
4. Measure the fluorescence of each sample at 450 nm.

One unit of enzyme activity is the amount of enzyme that catalyses the release of 1 nmol of 4-methylumbelliferone per hour.

[a] The assay of β-glucuronidase is performed exactly as for β-hexosaminidase, except that the substrate solution is 10 mM 4-methylumbelliferyl-β-D-glucuronide in 0.1 M sodium acetate, pH 4.8 (6).

Lysosomes are relatively dense organelles and are usually found at the bottom of the gradient. A second peak of lysosomal enzyme activity at a higher position in the gradient is often indicative of enzyme in vesicles derived from the endosomal compartment (see Section 3.1) and may also include enzyme located along the biosynthetic pathway (in the endoplasmic reticulum and/or the Golgi apparatus). A third peak may be found at the top of the gradient in a position corresponding to soluble proteins and may be indicative of lysis of lysosomes during the homogenization or fractionation procedures.

7. Inhibitors of receptor-mediated endocytosis and/or transport pathways

Much of the available information on the pathways followed by ligands and receptors following RME has come from studying the effects of inhibitors of various steps in the pathways. *Table 4* lists some of the major inhibitors of receptor-mediated endocytosis and/or subsequent transport pathways which have been found to be of use in experimental analysis of the pathways involved.

Table 4. Inhibitors of receptor-mediated endocytosis transport pathways and/or receptor/ligand processing

Inhibitor	References
Low temperatures	30–33
Hypertonicity of medium	34, 35
Potassium depletion of medium	36
Acidification of cytoplasm	37
Inhibitors of energy metabolism	2, 8
Weak base amines	3, 38
Ionophores	3, 38
Specific antibodies	39–41
Peptide proteinase inhibitors	38, 42

7.1 Inhibition of transport and degradation of ligands by low temperatures

As mentioned previously, receptor-mediated endocytosis is an energy-requiring process. While ligand binding by receptor occurs at 4 °C, no significant internalization of the resulting complex occurs below 10 °C. With increasing temperatures above 10 °C, the rate of internalization increases, reaching a maximum at 37 °C.

Another critical temperature is 20 °C. At this temperature, fusion of early endosomes (30), maturation of later endosomes (31), and the transport of ligand to a degradative (lysosomal) compartment is inhibited (32, 33). This observation has a useful practical implication. The incubation of cells with ligand at 20 °C may be a convenient means of loading up cells with ligand under conditions where it is not degraded. The cells can then be washed to remove unbound ligand and, on switching the cells to 37 °C, degradation generally occurs without a lag period (33).

7.2 Weak base amines and ionophores

The low pH which exists in the lumen of many components of the transport pathway involved in receptor-mediated endocytosis is an important regulator of the pathway. Agents such as weak base amines and monovalent carboxylic ionophores, which neutralize pH gradients, disturb these pathways.

Weak bases such as methylamine, ammonium chloride, and chloroquine diffuse readily into cells in their uncharged form. In intracellular acidic compartments they tend to become protonated and are unable to diffuse out. This trapping of protons within the lumen of endosomes and lysosomes increases the pH of the acidic compartment. The carboxylic ionophores such as monensin and nigericin dissipate intracellular pH gradients by a different mechanism: they allow protons to equilibrate across membranes by exchange with cations (3, 38).

The effects of lysosomotropic agents on the endocytic transport system can be very complex and include neutralization of acidic compartments, massive vacuolization of cells, inhibition of receptor–ligand dissociation (which often requires the low pH of these acidic compartments), inhibition of receptor recycling, and prevention of ligand degradation. *Figure 3* shows the inhibitory effect of ammonium chloride on the degradation of ^{125}I-EGF by fibroblasts.

Amines and ionophores have a rapid onset of effect, less than 30 sec. Their effects are readily reversed by wash-out. They seem to affect internalization to a lesser extent than recycling and so can be very useful practically to load cells with ligand at 37 °C without the occurrence of degradation. The following concentrations have been used: ammonium chloride, 10–50 mM; methylamine, 10 mM; chloroquine 25–200 μM; monensin, 50–100 μM.

It should be remembered that these reagents, chloroquine and the ionophores in particular, do have some side-effects in addition to their role in neutralizing pH gradients. The inhibition of lysosomal degradation, for example, may result in a reduced supply of amino acids especially over a prolonged long incubation period. In addition, other acidic compartments apart from those involved in the endocytic pathway may be affected. Monensin and nigericin are known primarily as inhibitors of Golgi functions, so they tend to have effects on Golgi-mediated transport in addition to their effects on acidic compartments.

7.3 Proteinase-inhibitory peptides

Specific protease inhibitors have been shown to block proteolysis in both endosomes and lysosomes. The use of such inhibitors has advantages over amines, ionophores, and 20 °C incubation in that they inhibit proteases *per se* rather than transport processes. In general, none of them are as effective as the amines but, because lysosomal enzymes tend to work sequentially in degradation, a single inhibitor can give quite a significant reduction in degradation (38). The author has found leupeptin (50 μg/ml) to be almost as effective as 10 mM ammonium chloride in inhibiting [^{125}I]EGF degradation in human fibroblasts. Inhibitors of lysosomal enzymes have been described in detail in a previous volume in this series (42). See also reference 38.

The inhibitors most often used include pepstatin, chymostatin, antipain, E64, leupeptin, and Z-PheAla-CHN_2. Of these antipain, E-64, and leupeptin are water-soluble (38). Leupeptin is a widely used inhibitor, it strongly inhibits the lysosomal cysteine proteases B, H, and L as well as many other proteases (42). It is water-soluble and relatively non-toxic, has been shown to be effective over a range 50–200 μg/ml, and is stored frozen as a 1 mg/ml stock solution. Some cells may require pre-incubation with the inhibitor for maximum effect.

The other inhibitors are only sparingly soluble in water and stock solutions must be made up in dimethyl sulphoxide (DMSO). These are only slowly taken up by cells by fluid-phase endocytosis and for maximum activity require pre-incubation with cells. A solvent control should always be included in the experiment. Pepstatin, an inhibitor of aspartic proteases, has been shown to be quite effective at very high concentrations (500 μg/ml) (18).

8. Determination of the rate of fluid-phase endocytosis

The rate of fluid-phase endocytosis of cells is determined by incubation of cells at 37 °C with a molecule which is impermeant to and does not adsorb to the plasma membrane. The amount of the molecule internalized in a given time is quantitated and used to estimate the rate of fluid-phase endocytosis. Various molecules have been used for this purpose, including horseradish peroxidase (HRP), radioiodinated species such as [^{125}I]BSA, [^{125}I]ovalbumin, and [^{125}I]poly(vinylpyrrolidone), [U-^{14}C]sucrose, and fluorescent markers such as lucifer yellow (43). Of these the most widely used is HRP (44, 45).

A true marker of fluid-phase endocytosis should not show any adsorption to the plasma membrane. In practice, most molecules stick to cell surfaces to some degree, so that determining the rate of fluid-phase uptake may be complicated by the existence of an adsorptive component. It is also important that all the extracellular component be removed so that the amount actually internalized can be determined accurately. Because fluid-phase uptake is an inefficient process, this can be of the order of 0.001% of that to which the cell is exposed. This means that exhaustive washing of the cells is necessary. The determination of the rate of fluid-phase endocytosis is discussed in detail in reference 43.

Protocol 7. Measurement of fluid-phase endocytosis

The washing procedure described here has been optimized for Hep 2 cells. Other cells may need more or less extensive washing.

1. Culture the cells in 35-mm Petri dishes.[a]
2. Wash the cells with 2.0 ml DMEM and then with 2.0 ml DMEM containing 5 mg/ml BSA.
3. Incubate the cells with DMEM containing 5 mg/ml BSA and 1–3 mg/ml HRP (Sigma, type II) for 2–6 h at 37 °C.[b] Incubate control dishes with HRP at 4 °C.
4. At the end of the incubation, rinse the cells three times with PBS containing 5 mg/ml BSA at room temperature. Incubate with 2 ml of DMEM, 5 mg/ml BSA for 10 min at 37 °C. Wash with PBS containing 5 mg/ml BSA, three 5-min washes at room temperature.
5. Solubilize the cells in 1.0 ml of 0.5% Triton X-100 at 4 °C.

6. Assay the cell extracts for HRP activity.

[a] Replicate dishes of cells should be carried through the protocol for estimation of cell numbers.

[b] With some cells, HRP can bind to the mannose receptor and can enter the cells through RME. This can be prevented by including yeast mannans (1–2 mg/ml; Sigma) in the uptake medium (46).

Protocol 8. HRP assay

Caution. *o*-Dianisidine hydrochloride is a potential carcinogen and should be handled with extreme care (gloves, face-mask).

Materials

- *o*-dianisidine HCl (Sigma)
- 0.3% H_2O_2
- 2% Triton X-100
- 0.5 M sodium phosphate, pH 5.0

Method

1. Prepare assay reagent[a] by mixing 12 ml of 0.5 M sodium phosphate, pH 5.0 with 6 ml of 2% Triton X-100 and 111 ml distilled water. Then add 13 mg of *o*-dianisidine HCl. Dissolve gently (no magnetic stirring) and then add 0.6 ml of 0.3% H_2O_2.
2. Place 0.2 ml of solubilized cells in a cuvette of 1.0 ml capacity.
3. Add 0.9 ml of the assay reagent. Mix by rapidly inverting the cuvette and monitor the optical density at 456 nm (OD_{456}).

Results

Rates of fluid-phase endocytosis are often expressed as the amount of HRP internalized (expressed as a percentage of the total HRP to which the cells are exposed)/unit time/mg of protein. However, it is preferable to relate the amount of HRP endocytosed per unit time to the number of cells present on the dish.

[a] The assay reagent should be made up fresh each day but, if necessary, can be stored in a dark bottle at 4 °C. It should not be used if it has a straw colour.

Acknowledgements

I would like to thank William S. Sly and Colin R. Hopkins in whose laboratories the experiments described in this chapter were performed. Also thanks to Jeff Grubb, Padhraig Nolan, and Maria Fiani for their help.

References

1. Wileman, T., Harding, C., and Stahl, P. (1985). *Biochem. J.* **232**, 1.
2. Gruenberg, J. and Howell, K. E. (1989). *Ann. Rev. Cell Biol.* **5**, 453.
3. Mellman, I., Fuchs, R., and Helenius, A. (1986). *Ann. Rev. Biochem.* **55**, 663.
4. Nolan, C. M., Kyle, J. W., Watanabe, H., and Sly, W. S. (1990). *Cell Regulation* **1**, 197.
5. Stahl, P., Schlesinger, P. H., Sigardson, E., Rodman, J. S., and Lee, Y. C. (1980). *Cell* **19**, 207.
6. Glazer, J. H. and Sly, W. S. (1973). *J. lab. clin. Med.* **82**, 969.
7. Lowry, O. H., Rosebrough, N. J., Farr, A. L., and Randall, R. J., (1951). *J. Biol. Chem.* **193**, 265.
8. Gonzalez-Noriega, A., Grubb, J. H., Talkad, V., and Sly, W. S. (1980). *J. Cell Biol.* **85**, 839.
9. Scatchard, G. (1949). *Ann. NY Acad Sci.* **51**, 660.
10. Limbird, L. E. (1986). *Cell surface receptors: a short course on theory and methods*. Martinus Nijhoff Publishing, Boston.
11. Harford, J., and Klausner, R. D. (1987). In *Methods in enzymology*, Vol. 149. (ed. R. Green and K. J. Widder), pp. 3–9. Academic Press, London.
12. Watanabe, H., Grubb, J. H., and Sly, W. S. (1990). *Proc. nat. Acad. Sci. USA* **87**, 8036.
13. Goldstein, J. L. and Brown, M. S. (1983). In *Methods in enzymology*, Vol. 98. (ed. S. Fleischer and B. Fleischer), pp. 241–60. Academic Press, London.
14. Haigler, H. T., Maxfield, F. R., Willingham, M. C., and Pastan, I. (1980). *J. Biol. Chem.* **255**, 1239.
15. Ascoli, M. (1982). *J. Biol. Chem.* **257**, 13306.
16. Ward, D. M., Ajioka, R., and Kaplan, J., (1989). *J. Biol. Chem.* **264**, 8164.
17. Breitfeld, P. P., Casanova, J. E., McKinnon, W. C., and Mostov, K. E. (1990). *J. Biol. Chem.* **265**, 13750.
18. Diment, S. and Stahl, P. (1985). *J. Biol. Chem.* **260**, 15311.
19. Schwartz, A. L., Fridovich, S. E., and Lodish, H. R. (1982). *J. Biol. Chem.* **257**, 4237.
20. Goldstein, J. L., Brown, M. S., Anderson, R. G. W., Russel, D. W., and Schneider, W. J. (1985). *Ann. Rev. Cell Biol.* **1**, 1.
21. Fischer, H. D., Gonzalez-Noriega, A., and Sly, W. S. (1980). *J. Biol. Chem.* **255**, 5069.
22. Wileman, T. E., Boshans, R. L., Schlesinger, P., and Stahl, P. (1984). *Biochem. J.* **220**, 665.
23. Morgan, D. O., Edman, J. C., Standring, D. N., Fried, V. A., Smith, M. C., Roth, R. A., and Rutter, W. J. (1987). *Nature* **329**, 301.
24. Hopkins, C. R. and Trowbridge, I. S. (1983). *J. Cell Biol.* **97**, 508.
25. Mostov, K. E. and Simister, N. E. (1985). *Cell* **43**, 389.
26. Merion, M. and Sly, W. S. (1983). *J. Cell Biol.* **96**, 644.
27. Sahagian, G. G. and Neufeld, E. F. (1983). *J. Biol. Chem.* **258**, 7121.
28. Aronson, N. N. and Touster, O. (1974). In *Methods in enzymology*, Vol. 31 (ed. S. Fleischer and L. Packer), pp. 90–102. Academic Press, London.

29. Fischer, H. D., Gonzalez-Noriega, A., Sly, W. S., and Morre, D. J. (1980). *J. Biol. Chem.* **255**, 9608.
30. Diaz, R., Mayorga, L., and Stahl, P. (1988). *J. Biol. Chem.* **263**, 6093.
31. Felder, S., Miller, K., Moehren, G., Ullrich, A., Schlessinger, J., and Hopkins, C. R. (1990). *Cell* **61**, 623.
32. Dunn, W. A., Hubbard, A. L., and Aronson, N. N. (1980). *J. Biol. Chem.* **255**, 5971.
33. Marsh, M., Bolzau, E., and Helenius, A. (1983). *Cell* **32**, 931.
34. Daukas, G. and Zigmond, S. H. (1985). *J. Cell Biol.* **101**, 1673.
35. Heuser, J. and Anderson, R. G. W. (1989). *J. Cell Biol.* **108**, 389.
36. Larkin, J. M., Brown, M. S., Goldstein, J. L., and Anderson, R. G. W. (1983). *Cell* **33**, 273.
37. Sandvig, K., Olsnes, S., Peterson, O. W., and van Deurs, B. (1987). *J. Cell Biol.* **105**, 679.
38. Seglen, P. O. (1983). In *Methods in enzymology*, Vol. 96 (ed. S. Fleischer and B. Fleischer), pp. 737–64. Academic Press, London.
39. Nolan, C. M., Creek, K. E., Grubb, J. H., and Sly, W. S. (1987). *J. Cell Biochem.* **35**, 137–51.
40. Anderson, R. G. W., Brown, M. S., Beisiegal, V., and Goldstein, J. L. (1982). *J. Cell Biol.* **93**, 523.
41. Schreiber, A. B., Lieberman, T. A., Lax, I., Yarden, Y., and Schlessinger, J. (1983). *J. Biol. Chem.* **258**, 846.
42. Gordon, P. B. and Seglen, P. O. (1989). In *Proteolytic enzymes: a practical approach* (ed. R. J. Beynon and J. S. Bond), pp. 201–10. IRL, Oxford.
43. Wiley, H. S. and McKinley, D. N. (1987). In *Methods in enzymology*, Vol. 196 (ed. D. Barnes and D. A. Sirbasku), pp. 402–17. Academic Press, London.
44. Steinman, R. M., Silver, J. M., and Cohn, Z. A. (1974). *J. Cell Biol.* **63**, 949.
45. Steinman, R. M. and Cohn, Z. A. (1972). *J. Cell Biol.* **55**, 186.
46. Rodman, J. S., Schlesinger, P., and Stahl, P. (1978). *FEBS Lett* **85**, 345.

2

Membrane traffic pathways in polarized epithelial cells

JULIET A. ELLIS, MARK R. JACKMAN, JORGE H. PEREZ, BARBARA M. MULLOCK, and J. PAUL LUZIO

1. Introduction

Epithelial cells maintain a selective permeability barrier between extracellular environments and regulate the composition of the fluids on either side of the cell layer. They achieve this by a diverse array of specialized activities including ion transport, secretion, and transcytosis (1), which are dependent on the polarized distribution of proteins and lipids between the apical and basolateral plasma membrane domains of the individual cells (2). Establishment and maintenance of cell surface polarity requires the formation of intercellular contacts and junctions, correct cell–substrate adhesion, highly ordered cytoskeletal networks, and specialized protein and lipid plasma membrane traffic pathways (3). This chapter confines itself to such membrane traffic pathways in polarized epithelia and, in particular, describes methods to examine the sorting of proteins to the correct plasma membrane domains. We provide protocols to study two different polarized cell types; a human, colon-derived, adenocarcinoma cell line with enterocyte-like morphology called Caco-2 (4), which can be grown on filters *in vitro* as a polarized cell layer, and rat hepatocytes (5), which can be studied *in situ* in rat liver. We also discuss data from the MDCK (Madin–Darby canine kidney), renal-cortex cell line (6), which can also grow in culture as a polarized cell layer and from which much of our present knowledge of membrane traffic pathways in epithelial cells is derived.

Experiments on protein sorting in polarized epithelial cells have revealed differences in the intracellular site of sorting of newly synthesized plasma membrane proteins, depending on cell type. Recent evidence suggests that protein sorting can occur not only in the *trans*-Golgi network (TGN), before vesicular transport to the cell surface, but also on the endocytic–transcytic route (*Figure 1*). In different epithelial cell types there is a different emphasis on the mechanism which dominates. In MDCK cells, proteins destined for the plasma membrane are sorted in the TGN into different constitutive, secretory

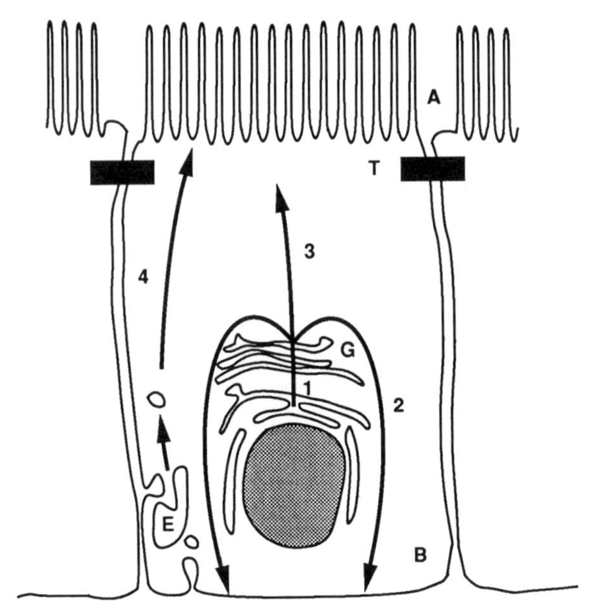

Figure 1. Routes of sorting of newly synthesized plasma membrane proteins. The diagram shows all known pathways for the delivery of newly synthesised membrane proteins to the cell surface. 1. Initial delivery from the endoplasmic reticulum to the Golgi complex (G). 2. Direct vectorial delivery to the basolateral surface (B). 3. Direct vectorial delivery to the apical surface (A). 4. The endocytic–transcytic route used by some apical proteins which are first delivered to the basolateral surface and then transcytosed via endosomes (E). T, tight junction.

vesicles and then targeted directly to the correct domain (3, 7). Newly synthesized secretory proteins are also released in a polarized manner from these cells (8). Hepatocytes, however, appear to rely entirely on endocytic–transcytic sorting of newly synthesized apical membrane proteins. All hepatocyte plasma membrane proteins so far studied are initially targeted from the TGN to the basolateral membrane. Proteins destined for the apical domain are then sorted at the basolateral membrane by endocytosis and are transcytosed to the apical domain (5). Secreted proteins leave hepatocytes exclusively through the basolateral membrane, suggesting that hepatocytes lack a direct exocytic route from the TGN to the apical surface and therefore need to sort plasma membrane proteins in a different way to MDCK cells. Caco-2 cells exhibit both types of transport pathway. For example, the brush border hydrolase sucrase-isomaltase is sorted in the TGN and transported directly to the apical domain (9, 10). In contrast, other brush border hydrolases such as dipeptidylpeptidase IV and aminopeptidase N reach the apical domain both by direct transport from the TGN and via transcytosis from the basolateral domain. As in hepatocytes, endogenous secretory proteins are released from the basolateral domain only.

It has been suggested that differential sorting of newly synthesized membrane proteins to the apical and basolateral domains occurs because only the apically directed pathway requires signal-mediated recognition, whereas proteins destined for the basolateral domain travel by default (11). To explain why some apical membrane proteins in Caco-2 travel to the basolateral membrane first, Simons and Wandinger-Ness (11) proposed that these proteins either lack the necessary apical sorting signal or escape selection, and therefore join the non-specific carrier vesicles targeted to the basolateral surface.

It should be noted that different methods have been employed to study plasma membrane protein transport in hepatocytes compared to that in other epithelial cell types. Subcellular fractionation has been the main technique used to study the apical and basolateral plasma membrane domains of hepatocytes, whereas cell-surface labelling techniques (e.g. domain-selective biotinylation/[^{125}I]streptavidin blotting) have been applied to the cultured epithelial cell lines. For direct comparison between different cell types, membrane protein sorting needs to be examined using the same techniques.

In this chapter, we describe methods to isolate the apical and basolateral plasma membrane domains from Caco-2 cells, methods for domain-specific cell-surface labelling, and a means of characterizing labelled proteins. The study of membrane traffic pathways in hepatocytes is restricted by the lack of well characterized, polarized, permanent cell lines, though some experiments are possible on primary cultured cells. However, an advantage of hepatocytes is the possibility of studying membrane traffic pathways while the cells form part of an intact organ. We provide protocols for establishing the isolated perfused rat liver as an experimental system and subcellular fractionation techniques for studying endocytic pathways in this organ *in situ*. We specifically exclude methods for preparing the blood sinusoidal (basolateral) and bile canalicular (apical) domains of rat hepatocyte plasma membrane since these are widely used and well described elsewhere (5, 12, 13).

2. Membrane traffic pathways in cultured Caco-2 cells

2.1 Cell culture

Caco-2 cells may be cultured in plastic flasks or as a truly polarized cell layer on a filter support (*Protocol 1*; *Figure 2*). The establishment and maintenance of epithelial polarity requires the presence of tight junctions which seal cells in an epithelial cell sheet, so that molecules of $M_r > 5000$ are prevented from leaking across. This is essential for the epithelium to become selectively permeable. Measurement of the 'tightness' of an epithelial monolayer can be made in several ways.

Protocol 1. Culture of Caco-2 cells

A. *Cell culture of Caco-2 cells in flasks*

1. Maintain Caco-2 cells in Dulbecco's modified Eagle's medium (DMEM; 3.7 g/l glucose) supplemented with 2 mM L-glutamine, 50 U/ml penicillin, 50 μg/ml streptomycin, 1% (v/v) non-essential amino acids, and 10% fetal calf serum (FCS). Cell cultures are grown at 37 °C in a humidified atmosphere.
2. Subculture the cells when they reach 80% confluency. Wash twice with phosphate-buffered saline (PBS) and detach the cells with 0.25% (w/v) trypsin and 0.125% (w/v) EDTA at 37 °C. Add an equal volume of DMEM. Centrifuge the cell suspension at 500 r.p.m. in a bench centrifuge for 5 min.
3. Carefully resuspend the cell pellet in a small volume (1 ml) of DMEM which is added dropwise while shaking the tube. One 75 cm^2 flask is split 1 in 5. A confluent 75 cm^2 flask contains 5×10^7 cells.

B. *Culture of polarized Caco-2 cell layers on membrane filters*

Caco-2 cells may be grown on permeable supports consisting of a polycarbonate filter in a plastic insert (0.45 μm pore, 24 mm diameter, collagen-coated, Transwell filters; Costar) which fits into a standard 6-well plate (supplied with filters). These allow Caco-2 cells to differentiate fully and easy access to apical and basolateral membrane surfaces.

1. Seed Caco-2 cells at a density of 2×10^6 cells per 2 ml DMEM to the upper chamber of each filter insert. Place the filters in a 6-well plate and add 2 ml DMEM to the lower chamber.
2. Allow the cells to adhere for 3 h, and then change the medium.
3. Subsequently, change the medium every 3 days. Cells become differentiated 5–7 days after confluence.

2.1.1 Measurement of transepithelial electrical resistance

Tight junction barrier function is often expressed in terms of resistance to the transepithelial passage of an electric current. Caco-2 cells grown on filters have electrolyte solutions on both the apical and basolateral sides. The measurement of electrical resistance across the monolayer is thus a measure of passive ion permeation. In practice this is measured using an epithelial voltohmmeter (Millicell-ERS, Millipore), which passes an alternating current across the monolayer and reads the resistance. Ohm's law is then applied to calculate resistance in ohms cm^2 monolayer surface. The electrical resistance across the monolayer steadily increases as the cells differentiate, reaching a plateau of ~ 500 ohm cm^2 when the cells are fully differentiated and a tight epithelium present.

2.1.2 Transport of water-soluble probes that are impermeable to the membrane

Impermeable membrane markers include Lucifer yellow CH, [^{14}C]polyethylene glycol, [^{14}C]mannitol, and [^{3}H]dextran (14). Markers to be avoided when examining the permeability properties of Caco-2 cells are [^{3}H]ouabain and [^{14}C]insulin, as both of these are affected by the physiology of the cell.

Protocol 2. Assessment of a Caco-2 epithelial monolayer using mannitol

1. Carry out transport at 37 °C in the absence of cells to assess the barrier properties of the filter, and then across cell monolayers of different ages in cell culture.[a] A time course of transport across the cell monolayer is necessary for each filter. Add [^{14}C]D-mannitol (60 mCi/mmol from Amersham) at the start to the apical surface ~ 50 000 d.p.m. in 1.5 ml of medium. The basolateral surface rests in marker-free DMEM.
2. At each sampling time, rapidly transfer the inserts to another well containing fresh transport medium.
3. Determine the amount of radionuclide transported at each time point in a liquid scintillation counter.

[a] Transport of the impermeable substrate should decrease with cell age, reaching a minimum of 0.5–1%, once the cells are fully differentiated.

2.2 Preparation of apical and basolateral plasma membrane fractions

Subcellular fractionation techniques have been successfully applied to a variety of intact tissues including rat liver, colon, and ileum (15–17) but have proven frustratingly difficult when applied to cultured cells (18). The difficulties include the following.

- The small amount of starting material available and the long times taken to culture sufficient numbers of cells.
- Retention of cytoplasmic organization after standard homogenization procedures. This causes aggregation and loss of organelles and plasma membrane vesicles in early centrifugation steps.
- The inability to release all the organelles as separate entities into suspension (because of the above).

- The large variation in the extent of the cytoplasmic organization between different cell culture types. Thus, homogenization conditions have to be standardized for each cell type.

In *Protocol 3* a method for the simultaneous purification of apical and basolateral plasma membrane fractions from Caco-2 cells is described. Enzyme markers used to assess the purity of fractions were sucrase-isomaltase for the apical membrane and Na^+K^+ATPase for the basolateral membrane. Data from a representative purification are shown in *Table 1*.

Protocol 3. Isolation of apical and basolateral domains

Brush border and basolateral membrane vesicles from filter-grown Caco-2 cells are isolated by sucrose density gradient centrifugation and differential precipitation adapted from methods previously described (19, 20).

1. Rinse filter-grown Caco-2 cells once in ice-cold 0.9% (w/v) NaCl, once in buffer A[a] and scrape off with a rubber policeman. Suspend the cells in 5 ml buffer A. All further operations are carried out at 4 °C.
2. Equilibrate the cells with N_2 for 10 min at 550 lb/in^2 (17) in a cell disruption bomb (Type 4639, Parr Instrument Co.). Carefully open the discharge valve allowing the sample to come out dropwise into a collecting vessel. Microscopic examination of the homogenate showed ~ 95% cell breakage.
3. Degas the homogenate for 15 min and then centrifuge for 10 min at 270 *g* in a SS34 rotor, (Sorvall Instruments Division) in a Sorvall RC-5 centrifuge.
4. Collect the resulting supernatant (SN1) and centrifuge at 920 *g* for 10 min in the same rotor.
5. Collect the supernatant (SN2), add $MgCl_2$ to 10 mM, and stir for 15 min on ice.
6. Centrifuge SN2 for 15 min at 2300 *g* in a SS-34 rotor. Pellet the resulting supernatant (SN3) at 170 300 *g* for 45 min in a T-1270 rotor (Sorvall Instruments Division) in a Sorvall OTD 65B centrifuge.
7. Resuspend the crude membrane pellet (P4) in 0.5 ml buffer B[b] and layer on the following discontinuous sucrose gradient (w/w): 1.41 ml, 45%; 1.64 ml, 35%; 1.42 ml, 30% in a 13 × 51 mm Beckman polyallomer tube. All the sucrose solutions are buffered with 10 mM Tris-HCl pH 7.4 and their density checked with a refractometer (Bellingham and Stanley Ltd).
8. Centrifuge for 4.5 h in a Beckman SW 50 swinging bucket rotor in a Beckman L5-50B ultra centrifuge at 68 000 *g*. This gradient can also be centrifuged in a vertical rotor (Sorvall, TV-865) at 219 000 *g* for 1.5 h.

9. Collect the bands at the interfaces (designated as I: overlay/30%; II: 30%/35%; III: 35%/45%) with a syringe and needle and dilute to 8% with buffer A. Add $MgCl_2$ to fraction II (final concentration, 10 mM). After gentle agitation on ice for 15 min centrifuge the fraction for 15 min at 2300*g* in an SS-34 rotor.

10. Separately pellet the supernatant (SN5, from fraction II) and fractions I and III at 36 000 *g* for 1 h in the SS-34 rotor. Resuspend the pelleted fractions I (Golgi-enriched), II (basolateral plasma membrane) and III (apical plasma membrane) in a minimal volume of buffer A.

Enzyme assays

The following marker enzymes are used: sucrase-isomaltase (21) for brush border (apical) membranes; K^+-stimulated *p*-nitrophenyl-phosphatase (Na^+K^+ ATPase; 20) for basolateral membranes; *N*-acetyl-*ß*-glucosaminidase (22) for lysosomes; neutral α-glucosidase (23) for endoplasmic reticulum; UDP-galactosyl transferase (uridine diphosphate galactosyl transferase) (24) for the Golgi complex; and succinate dehydrogenase (25) for mitochondria.

[a] Buffer A: 0.25 M sucrose, 12 mM Tris, pH 7.4 with HCl.
[b] Buffer B: 0.25 M sucrose, 5 mM EDTA, 12 mM Tris, pH 7.4 with HCl.

Table 1. Recovery of marker enzymes in purified plasma membrane fraction from Caco-2 cells

	Enrichment x, relative to homogenate		**Yield (% homogenate)**	
Marker enzyme	**Apical**	**Basolateral**	**Apical**	**Basolateral**
Sucrase isomaltase	14.3	2.0	13.0	0.95
Na^+K^+ATPase	1.3	20.0	1.1	13.0

The marker enzymes *N*-acetyl-*ß*-glucosaminidase, neutral α-glucosidase, UDP-galactosyl transferase and succinate dehydrogenase were all depleted in both apical and basolateral membrane fractions being present at $< 0.6\%$ of homogenate values.

2.3 Biosynthesis of plasma membrane proteins

The subcellular fractionation method described above for the simultaneous isolation of apical and basolateral membranes from a polarized epithelial cell line provides a method for examining the polarized expression of surface membrane proteins in cell lines analogous to that used for hepatocytes (5). *Protocol 4* gives a method for pulse-chase labelling and immunoprecipitation of membrane proteins.

Protocol 4. Biosynthetic labelling and immunoprecipitation of membrane proteins

Grow Caco-2 cells on filters as described in *Protocol 1* and use for labelling between 17 and 21 days after seeding.

1. Wash cell monolayers in cysteine- and methionine-free DMEM (CMFM) for 30 min at 37 °C in a 7% CO_2 humidified incubator (1 ml CMFM in both upper and lower chambers).
2. Invert the filters and apply the label to the basolateral surface. The labelling reagent recommended is Trans ^{35}S-label TM (ICN Radiochemicals) which contains 70% L-methionine (^{35}S) and 15% L-cysteine (^{35}S) and is diluted into CMFM. Add 200 μCi (in a volume of 200 μl/filter) to each filter for 15–30 min at 37 °C. (The long pulse time is to overcome the relatively large non-specific binding of the label to the filter.)
3. At the end of the pulse time, wash the filters in DMEM to remove excess label. Chase in DMEM, supplemented with 10 mM cysteine and 10 mM methionine, for the appropriate times at 37 °C. The filters are now used upright, and 1 ml of chase medium is added to both the apical and basolateral chambers.
4. At the end of the chase times, and for time O, 'dunk' the filters into ice-cold buffer A (see *Protocol 3*) and scrape the cells from the filter into buffer A. About 1 ml per filter of buffer A is required. Normally three filters are used per time point, and carrier cells are added (1/2 of a 150 cm^2 confluent flask's worth/timepoint).
5. Homogenize whole cells from each time point and prepare the apical and basolateral membrane fractions as described in *Protocol 3*.
6. Immunoprecipitate the apical and basolateral membranes (using the same amount of protein for each; typically 100 μg) with an appropriate rabbit antiserum or IgG fraction. Resuspend the membrane fractions from the gradient in TNS[a] and add Triton X100 (TX100) to 2%. Whirlymix samples, microcentrifuge for 10 min at ~ 12 000 *g*, and discard any insoluble pellet.
7. Carry out a pre-incubation for each sample to reduce non-specific binding to the Protein A–Sepharose used for solid-phase immunoprecipitation. Mix each sample with 1 μg non-immune rabbit IgG and 20μl Protein A–Sepharose (Sigma; 50 mg/ml slurry in phosphate-buffered saline (PBS[b]) made up fresh), whirlymix, and rotate for 2 h at room temperature.
8. Remove Protein A–Sepharose by microcentrifuging for 20 min at ~ 12 000 *g*, room temperature and transfer the supernatants to clean Eppendorf microcentrifuge tubes.

9. Add the rabbit antibody of interest to each sample at the appropriate dilution and incubate overnight at 4 °C with rotation.
10. Again add Protein A–Sepharose to the samples (this time to bind to the Fc regions of the first antibody). The amount added needs to be an excess over the presence of IgG. If the first antiserum is added at 1:500 dilution (1 μl to 0.5 ml sample) and serum is approximately 10 mg/ml IgG, then in 1 μl there is 10 μg IgG. Protein A-Sepharose is added so that it is at a final concentration of ~ 5 times its IgG binding capacity. 12.5 μl of a 50mg/ml slurry (capable of binding 0.625 mg IgG) is added to each sample. Incubate for 2 h at room temperature with rotation.
11. Microcentrifuge samples, for 20 min at ~ 12 000 *g* at room temperature and discard the supernatants. Beware of the fluffy pellets!
12. Wash the Sepharose pellets in TNSX[c] five times. Rotate for 10 min between each wash.
13. Wash two times in 50 mM Tris-HCl pH 7.4.
14. Drain the pellet, and add SDS–PAGE sample buffer.
15. Heat the samples to 95 °C for 5 min, cool, and lightly microcentrifuge. Run a standard 10% SDS–PAGE gel (26); then stain and destain it.
16. Fluorograph the gel using NEN 'EN^{3}HANCE™' according to the maker's instructions. Dry the gel down and put it up as an autoradiograph using KODAK X-OMAT™ AR film.

Samples can also be immunoprecipitated using monoclonal antibodies, in which case sheep anti-mouse IgG Dynabeads™ M-450 (DYNAL) are used in place of protein A–Sepharose (which does not bind well to mouse, sheep, or rat IgG at pH 7.4). Dynabeads are supplied as a suspension of 4×10^8 beads/ml (30 mg/ml) in PBS, and washed well before use. 1 mg Dynabeads is used with 0.2–2.0 μg monoclonal antibody.

[a] TNS: 50 mM Tris, 100 mM NaCl, 0.4% SDS, pH 7.4 with HCl.
[b] PBS: 0.15 M NaCl, 2 mM NaH_2PO_4, 16 mM Na_2HPO_4, pH 7.4.
[c] TNSX: 50 mM Tris, 100 mM NaCl, 0.1% SDS, 0.5% TX100, pH 7.4 with HCl.

2.4 Domain-specific cell-surface labelling techniques

The identification of domain-specific plasma membrane proteins by subcellular fractionation suffers the serious disadvantage that plasma membrane fractions are invariably contaminated by significant amounts of endomembranes (for example from the endoplasmic reticulum or the Golgi complex), as well as other plasma membrane domains. Cell-surface labelling of intact cells negates these problems and allows domain-specific labelling if the cells are grown as an integral epithelial monolayer on a filter support, thus confining the reactive labelling reagent solely to the basolateral or apical medium compartment (*Figure 2*). The use of polycarbonate filters is advisable since the high non-specific binding capacity of nitrocellulose filters is avoided (27).

Cell-surface labelling requires the labelling reagent to react exclusively at the cell surface with no intracellular labelling. Thus, such reagents must be membrane-impermeant (a property conferred upon the labelling reagent by its charged and hydrophilic nature) and sufficiently reactive at 4 °C, a temperature at which endocytosis is negligible. Labelling methods must be mild so as not to cause cell lysis. The duration of labelling and concentration of labelling reagent must be optimized for each cell type and system.

A wide variety of reagents are available commercially for cell-surface labelling, varying in their reactive groups (a comprehensive list of these are given in reference 28), and several may be used to label specifically one plasma membrane domain prior to examining membrane traffic pathways delivering labelled proteins to intracellular sites or to the opposite cell surface.

2.4.1 Cell surface iodination

Specific labelling with radioactive iodine is an important technique used to identify proteins that are exposed on the external surface of cell membranes. In labelling ^{125}I is preferred over ^{131}I since its γ-rays are less penetrating, making it safer to use, and its half-life is longer (60 days compared to 8 days). Iodination requires the formation of the $^{125}I^{+}$ cation from the oxidation of carrier-free Na [^{125}I]. Such oxidation can be catalysed by impermeant reagents either chemically (e.g. with Iodogen; references 29 and 30), or enzymatically (e.g. with lactoperoxidase; references 30–2). Incorporation of label reflects the presence of available membrane proteins with exposed tyrosine residues (and histidines in more alkaline conditions). It must also be realized that most iodinated proteins are likely to have unusually large extracellular domains (33). Despite the extensive use of radioiodination to

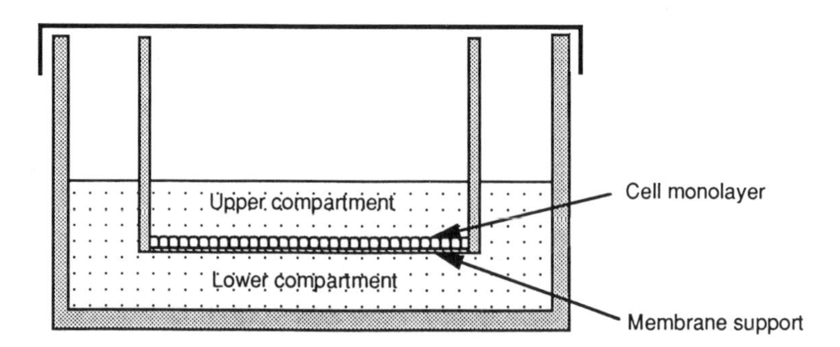

Figure 2. Polarized epithelial cells growing on a filter support. See *Protocol 1* for details of chamber dimensions and method of culture.

label cell-surface membrane proteins, very little use of this method has been made to label such proteins in polarized cell lines. Lactoperoxidase-catalysed iodination has been used to investigate the protein composition of the plasma membrane domains of MDCK cells. However, the interpretation of results using this procedure is made difficult because of the poor accessibility of the enzyme through the filter support (27). Apical membrane proteins of Caco-2 cells have been labelled using Iodogen-coated coverslips in this laboratory, but the method has proved unsuitable for labelling basolateral membrane proteins. Although radioactive labelling by iodination allows identification of cell-surface proteins, it does not permit their isolation. The use of biotin as a cell-surface label allows identification and isolation of membrane proteins.

2.4.2 Cell surface biotinylation

Domain-specific biotinylation of membrane proteins has been extensively used in studies of polarized cell lines (34–8). Derivatization with sulpho-NHS-biotin (sulphosuccinimidobiotin, the water-soluble analogue of NHS-biotin) requires a free amino group, and is therefore specific for membrane proteins with either extracellular unblocked N-terminal amino acids and/or extracellular free lysine residues, as well as amino lipids/sugars (*Figure 3*). Since lysine is a more abundant protein constituent than tyrosine a wider variety of proteins can be detected using sulpho-NHS-biotin than with iodination. Sulpho-NHS-biotin carries an overall net negative charge preventing its diffusion across cell membranes. Because it has a low molecular weight it can traverse a filter support and therefore efficiently label the basolateral surface of polarized epithelial cells. The labelling reaction can proceed at 4 °C, at close to physiological pH.

Biotin binds to streptavidin with one of the strongest non-covalent associations known (K_d approx 10^{-15} M). This allows identification of

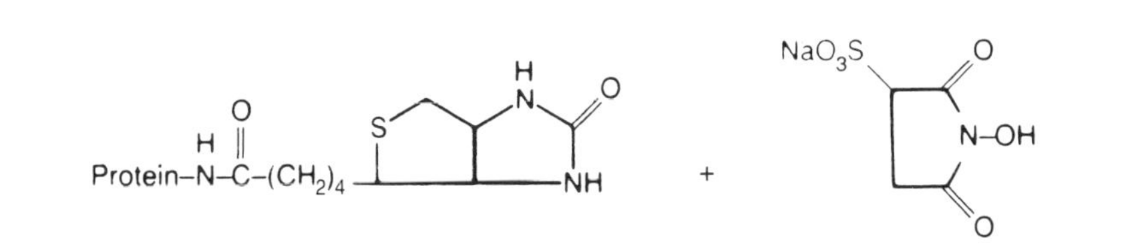

Figure 3. The reaction of sulpho-NHS-biotin with protein amines.

biotinylated proteins (by electrophoresis followed by electroblotting and detection with ^{125}I-streptavidin) down to picogram amounts under high stringency conditions (39). This labelling technique is considerably more sensitive than lactoperoxidase-catalysed cell surface iodination when applied to MDCK cells (27). Biotinylation also offers the advantage that biotinylated proteins can be visualized at the light and electron microscope level with an appropriate streptavidin conjugate. There are many biotinylation reagents available commercially for labelling cell surface molecules (e.g. from Pierce and Calbiochem), which can be tailored to suit the individual system and its requirements. A method for domain-specific labelling of Caco-2 cells is provided in *Protocol 5*. *Protocol 5* can be adapted to determine the polarity of a given antigen. Cells are labelled as described but are lysed in a suitable immunoprecipitation buffer (e.g. 1% Nonidet P40, 0.4% sodium deoxycholate, 66 mM EDTA, 10 mM Tris, pH 7.4 with HCl) and immunoprecipitated with the desired antiserum. Non-specific binding of [^{125}I] streptavidin to IgG heavy chain can be overcome by coupling affinity-purified antibodies to Sepharose beads and using a one-step immunoprecipitation procedure, or by including 1% (w/v) dry milk in the BSA–TGG buffer during the initial blocking step.

Protocol 5. Biotinylation of the apical and basolateral surfaces of Caco-2 cells

1. Wash a confluent cell monolayer grown on a polycarbonate filter support four times with PBS/CM[a]. Buffers containing amines (i.e. Tris, azide, glycine, ammonia) should be avoided as these may compete with the ligand during labelling reactions.
2. Thaw immediately before use an aliquot of frozen (−20 °C) stock NHS–LC–biotin (Pierce) solution (200 mg/ml in dimethylsulphoxide) and dilute to a final concentration of 1–1.5 mg/ml in cold PBS/CM.
3. For apical biotinylation, add 250 μl of NHS–LC–biotin solution to the apical chamber. Transfer the filter to a six-well plate filled with 1.5 ml cold PBS/CM and incubate on a rocker platform for 20–30 min at 4 °C.

   For basolateral biotinylation add 1 ml of cold PBS/CM to the apical compartment and place the filter chamber on a 250 μl drop of NHS–LC–biotin resting on a piece of Parafilm. This is then incubated for 20–30 min at 4 °C.
4. Wash the filters four times with cold PBS/CM.
5. To visualize biotinylated proteins, excise the filters from the chamber with a scalpel and extract with 1 ml of ice cold lysis buffer[b] for 1 h. Clarify the samples by centrifugation (~ 12 000 *g* for 10 min at 4 °C) and collect the supernatants. NB At this point biotinylated proteins may be separated into hydrophilic 'peripheral' membrane proteins and hydro-

phobic 'integral' membrane proteins by phase separation with Triton X-114 (TX-114) detergent as described by Bordier (40). However, it must be noted that some membrane proteins of intermediate hydrophobicity are found to partition with the aqueous phase (41) (see also Chapter 10).

6. Precipitate aqueous or detergent TX-114 phase cell extracts with five volumes of acetone (−20 °C, 30 min), centrifuge (~ 12 000 *g*, 10 min), and resuspend in Laemmli (26) sample buffer. Heat at 95 °C for 2–3 min, and separate by SDS–PAGE under reducing conditions (26). Separated proteins are then electroblotted on to nitrocellulose as described by Towbin *et al.* (42).
7. Block the nitrocellulose sheets with BSA-TGG[c] for 1 h at room temperature. Add [^{125}I]streptavidin (radiolabelled with the iodogen method (29) to a specific activity of 5 μ Ci/μg, 1–2 × 10^6 c.p.m./ml in TGG–0.3% BSA) for 2 h at room temperature. Wash the nitrocellulose four times (15 min each) in PBS–0.5% Tween 20.
8. Dry the blots and autoradiograph (2–16 h, −70 °C, with an intensifying screen) on Kodak XAR-5 film.

[a] PBS/CM:PBS (see *Protocol 4*) supplemented with 1 mM $MgCl_2$ and 0.1 mM $CaCl_2$.
[b] Lysis buffer: 0.15 M NaCl, 1 mM EDTA, 1% TX114, 10 mM Tris, pH 7.4 with HCl (protease inhibitors may be added if desired).
[c] BSA–TGG: 3% BSA (bovine serum albumin) in PBS containing 0.5% Tween 20 (Sigma), 10% glycerol, 1 M glucose.

Figure 4 shows the differential labelling of membrane proteins in each domain. It should be noted that other biotinylation reagents are reactive towards thiol groups (iodoacetyl–LC–biotin, Pierce; Maleimidobutyryl-Biocytin, Calbiochem) and biotin hydrazides will react with oxidized carbohydrate moieties of membrane glycoproteins and glycolipids.

2.4.3 Antibodies to cell-surface proteins

Antibodies raised to membrane proteins of either the apical or basolateral domain have proven useful in studying the biosynthetic pathways of such proteins in polarized cells (see Section 2.3) and may also be used to follow endocytic pathways. However, there are many technical problems in the domain-specific antibody labelling of cell-surface glycoproteins under physiological conditions. Apart from the problems of raising antibodies to membrane glycoproteins of cells grown in tissue culture (due to the small amounts of antigen usually available, and the strong possibility of raising antibodies to common non-specific sugar moieties of glycoproteins and glycolipids present in large amounts), the filter support offers a significant accessibility barrier to the diffusion of antibodies to the basolateral membrane surface. In addition, multivalent ligands, such as antibodies (and

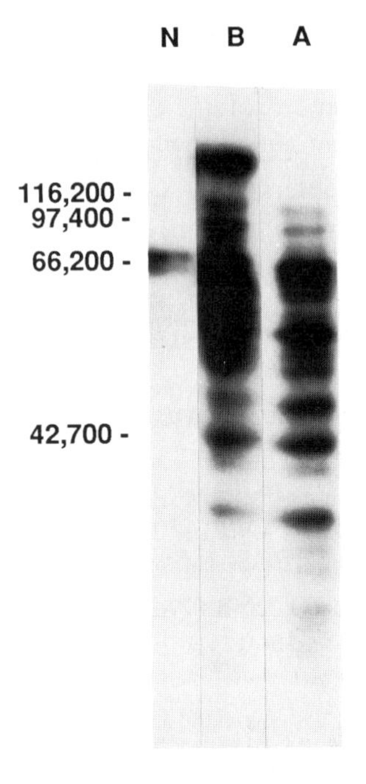

Figure 4. Biotinylated apical and basolateral plasma membrane proteins from Caco-2 cells. After domain-specific biotinylation-labelled membrane proteins were separated by SDS–PAGE, transferred to nitrocellulose and incubated with [^{125}I]streptavidin (see *Protocol 5*). N, non-specific binding of [^{125}I]streptavidin; A, biotinylation of the apical membrane; B, biotinylation of the basolateral membrane.

also lectins), binding to specific membrane proteins tend to cross-link them into large clusters forming 'patches'. Antibodies directed to membrane proteins may increase their rate of degradation, e.g. some monoclonal antibodes raised against the insulin receptor accelerate its rate of degradation by two orders of magnitude more than that caused by the natural ligand, insulin (43).

Antibodies directed to membrane proteins have also been used for indirect radioimmunoassay of apical or basolateral cell surface antigens of polarized cells grown on filter supports. The problem of antibody accessibility to the basolateral surface through the filter was reduced in one study by the use of Fab′ fragments of antibodies as opposed to the whole antibody (44).

2.4.4 Cell-surface carbohydrate (glycoprotein) labelling

Externally exposed glycoproteins can be radiolabelled by selective oxidation of the appropriate carbohydrate residue to form a reactive carbonyl on the membrane. Thus, sialic acid can be selectively oxidized by mild treatment with low concentrations ($\leq$ 1 mM) of periodate at 0 °C for short reaction times, so that the periodate is membrane-impermeable. Similarly, galactose can be selectively oxidized by galactose oxidase. The newly formed carbonyl groups can then be either:

- reacted with a fluorescent amine, and fixed by permanent reduction with sodium borohydride (45), or;
- reduced with tritiated borohydride (46; *Protocol 6*).

Protocol 6. Radiolabelling terminal sialic acid residues on apical and basolateral glycoproteins/glycolipids of Caco-2 cells

1. Wash the intact cell monolayer grown on a polycarbonate filter three times with PBS/CM (see *Protocol 5*).
2. For apical oxidation of sialoconjugates dissolve 1 ml of $NaIO_4$ in PBS/CM and add to the apical side of the filter. Transfer the filter to a six-well dish filled with 2 ml of ice-cold PBS/CM.
   For basolateral oxidation of sialoglyconjugates add 2 ml of 1 mM $NaIO_4$ in PBS/CM to the basolateral side of the filter. Add 1 ml of ice-cold PBS/CM to the apical side of the filter.
3. Incubate cells on ice for 30 min.
4. Wash monolayers four times with PBS/CM.
5. Add 1–5 mCi of $NaB[^3H]_4$ (350 mCi/mmol obtained from Amersham) in 125 µl PBS, pH 8.0 (prepared from stock solution of $NaB[^3H]_4$ (200 mCi/ml) in 0.01 M NaOH, stored at −70 °C) to the oxidized cell surface.
6. Incubate cells on ice in a humid chamber for 30 min and then wash three times with ice-cold PBS/CM.

Another common approach is to radiolabel externally exposed carbohydrate residues by enzymatically adding a radiolabelled monosaccharide, e.g. exogalactosylation (*Protocol 7*).

Protocol 7. Exogalactosylation of apical and basolateral glycolipids/ glycoproteins of ricin-resistant Caco-2 cells

1. Wash the intact cell monolayer grown on a polycarbonate filter twice with PBS/CM (see *Protocol 5*) and once with Hepes buffer[a].
2. For apical exogalactosylation, add 100 µl of reaction mixture[b] to the apical side of the filter. Transfer the filter to a six-well dish filled with 1.5 ml PBS/CM, and incubate on a rocking platform for 40–60 min at 4 °C.
   For basolateral exogalactosylation, add 1 ml of PBS/CM to the apical side of the filter, and then place the filter on a 100-µl drop of the reaction mixture on Parafilm.

Protocol 7. *Continued*

3. Stop the reactions by removing the reaction mixture and washing the filter three times with PBS/CM.

[a] Hepes buffer: 10 mM Hepes, 150 mM NaCl, 0.1% (w/v) BSA, pH 7.3 with HCl.
[b] Reaction mixture: Hepes buffer supplemented with 10 mM $MnCl_2$, 0.25 U/ml galactosyltransferase and 80 μCi/ml UDP-D-[6-^{3}H] galactose (18 Ci/mmol obtained from Amersham).

It is also possible to radiolabel terminal galactosyl residues using sialytransferase and radiolabelled CMP-sialic acid (47). Brandli *et al.* (48) have utilized a ricin (RCAII)-resistant MDCK cell line, deficient in cell surface galactose residues, to label membrane glycoproteins by exogalactosylation using exogenously added galactosyltransferase and UDP-galactose. Lectin-resistant cell lines exhibiting impaired galactosylation of glycoconjugates are particularly amenable to surface labelling by the above technique, as they allow maximal incorporation of label, but one must bear in mind that accessibility of the exogenously added enzyme may be limited.

2.5 Analysis of cell-surface labelling by two-dimensional polyacrylamide gel electrophoresis

In membranes it is commonly found that a few proteins make up a large percentage of the total; thus it is desirable to load the maximum possible amount of protein on to the resolving gel in order to visualize as many individual species as possible. Much larger amounts of protein can be applied to an isoelectric focusing gel without running into overloading problems than can be applied to sodium dodecyl sulphate polyacrylamide gel electrophoresis (SDS–PAGE). Since many plasma membrane proteins are labelled by most labelling methods, the high resolution of two-dimensional isoelectric focusing PAGE (2D IEF-PAGE) offers better identification of labelled proteins than SDS–PAGE. Caution must be used in identifying minor spots or those not regularly observed as surface proteins, since they could represent transient species secreted by the cell, or proteins expressed by any lysed cells in the culture. Many labelling procedures will alter the charge on modified proteins and thus change their isoelectric migration, e.g. by the loss of a protonated amino group. When this occurs, affected proteins must ideally be labelled uniformly throughout their populations to avoid the problem of artefactual charge heterogeneity of identical protein molecules, leading to multiple spots separating in the isoelectric focusing direction, but running with the same molecular weights in the second dimension.

For efficient 2D IEF-PAGE of labelled membrane proteins they must be completely solubilized. Anionic detergents such as SDS and sodium cholate have proven capable of completely solubilizing membrane proteins (49); however, these detergents obviously modify the charge properties of proteins to which they are bound and are therefore, on their own, not well suited for use in charge fractionation methods. Membrane proteins can be solubilized by SDS when followed by displacement of the SDS by high concentrations of urea and a non ionic detergent (50). Non-ionic (Triton X-100 and the closely related Nonidet P-40) and zwitterionic detergents (CHAPS and SB14) are capable of preserving the surface net charge of solubilized proteins and thus are able to overcome the above problem and have been used successfully in solubilizing membrane proteins (51). Work in the authors' laboratory has shown that solubilization of labelled Caco-2 cell membrane proteins by SDS followed by removal of SDS using excess (10-fold) Nonidet P-40 is adequate for resolution of labelled membrane proteins on 2D IEF-PAGE using the method of O'Farrell (52). The lysis buffer used was 0.2% w/v SDS; 2% Nonidet P-40; 9.8 M urea; 1.5% ampholines (pH 4–6.5, Pharmalyte; pH 3–10, Servalyte; and pH 5–7, LKB); 100 mM dithiothreitol. To resolve the cell-surface labelled membrane proteins of MDCK cells the following lysis buffer in conjunction with 2D IEF-PAGE as described by Bravo (53) has been shown to be efficient: 4% w/v NP40; 9.8 M urea; 2% ampholines (pH 7–9; LKB); 100 mM dithiothreitol (54).

2.6 Investigation of endocytic and transcytic pathways

Endocytic pathways may be studied as in other cells by the addition of exogenous labelled ligands. Thus, the internalization and recycling of transferrin has been studied in Caco-2 cells by pulse-chase addition of [^{125}I]transferrin to the basolateral chamber followed by sampling the basolateral chamber for the reappearance of counts after different times of incubation at 37 °C. The lack of counts appearing in the apical chamber eliminated the possibility of transcytosis of this ligand in Caco-2 cells (55). Similarly, transcytosis in Caco-2 cells has been followed using radioactive cobalamin (56).

Access to both apical and basolateral compartments by the growth of a polarized epithelium on filter supports has enabled the identification of a common endosomal compartment which can be entered from both the apical and basolateral surfaces. This has been shown, using confocal microscopy and electron microscopy, to follow the endocytic pathways of distinguishable markers applied to either the apical or basolateral cell surface (55, 57).

The ability to achieve domain-specific cell-surface labelling and analysis of proteins labelled has allowed endocytic recycling routes and transcytic routes to be analysed. Elegant studies by Brandli and Simons (54) using MDCK II cells resistant to the lectin toxin ricin have investigated cell-surface glycoproteins that recycle between the TGN and either the apical or the basolateral membranes and also allowed identification of glycoproteins that are transcytosed between the two cell-surface domains. The ricin-resistant MDCK II cells used are deficient in translocation of UDP-galactose into the Golgi complex. Consequently, surface glycoproteins expressed in these cells lack N-linked galactose residues and the terminal sialic acids that would otherwise be attached to these galactose residues. This defect enables surface glycoproteins of either membrane domain to be efficiently exogenously labelled with [^{3}H]galactose. This covalent modification generates substrates for sialytransferase present in the TGN. Thus, any sialic acid containing glycoproteins present on a particular membrane domain after exogenous galactosylation would have been endocytosed, transported to the TGN, and recycled back to that membrane domain after sialylation. Any such glycoproteins can be detected by specific lectin-binding to the sialic acid residues or by oxidation followed by reduction with NaB[^{3}H]$_4$ and analysis of extracted membrane by 2D IEF-PAGE. This approach has identified a restricted set of apical proteins that recycle through the TGN in MDCK II cells (54). Similarly, the transcytosis of a restricted set of exogenously [^{3}H]galactosylated glycoproteins was detected by surface biotinylation of glycoproteins of the opposite membrane domains, detergent solubilization of these glycoproteins, absorption on to streptavidin–agarose, and, finally, separation by 2D IEF-PAGE (58).

Other approaches to studying transcytic routes in polarized cells have included:

- using sulpho-NHS-SS-biotin to label one membrane domain followed by measuring the fraction of this probe that becomes resistant to glutathione added to the labelled domain, and accessible to glutathione added to the opposite domain during incubation at 37 °C;
- studying MDCK II cells transfected with isoforms of the Fc receptor (59).

3. Membrane traffic pathways in rat hepatocytes

3.1 Primary culture of hepatocytes

Primary culture of rat hepatocytes has been described previously in the *Practical Approach* series (60). The cells may be grown on plastic- or collagen-coated supports (61) and form a near-confluent monolayer with numerous phase bright spaces forming at cell/cell interfaces after 24–36 h in culture. These spaces are bile canalicular structures and can concentrate exogenously added fluorescein and pIgA. Unfortunately, it is difficult to gain access to these spaces to add reagents or sample them. Thus, whilst the cells regain polarity in cell culture, they are of limited use in studying membrane traffic pathways. Some microscopy studies have been carried out after exogenous addition of ligands for endocytosis as well as biosynthetic pulse-chase labelling studies of membrane proteins known to be concentrated at specific cell surface domains (62, 63). For such studies the techniques used as similar to those described in *Protocol 4* for Caco-2 cells.

3.2 Endocytic and transcytic pathways of hepatocytes *in situ*

Whilst there are limitations to studying membrane traffic pathways in cultured hepatocytes compared to Caco-2 cells, it is easier to study such pathways *in situ* in liver than in gut. In intact liver, endocytic and transcytic pathways can only be followed by monitoring the movement of suitably labelled ligands at increasing times after introduction of a pulse of ligand into the system. In the whole animal, introduction usually involves intravenous injection. A ligand appearing intact in the bile 15–20 min later, e.g. polymeric IgA, can be used as a marker for the transcytic pathway; ligands which are broken down by the hepatocyte, e.g. asialoglycoproteins, serve as markers for the endocytic pathway leading to lysosomes. There is a wealth of evidence that, following receptor-mediated endocytosis at the blood sinusoidal cell surface, internalized ligands and their receptors are sorted in an early endosome compartment and probably move from this compartment to their eventual destination by vesicular transport (64–66). In practice, liver endocytosis experiments in the whole animal are very difficult to control, since other organs may remove or modify part of the material intended for the liver, whether ligand or inhibitor. It is also difficult to achieve a true pulse dose; material failing to be taken up by the liver on the first pass may well be taken up on recirculation. An isolated perfused liver system can be controlled and examined much more readily. It is also advantageous in that it enables the investigation of the effects of one ligand on the endocytosis of another and makes possible the quantification of the relative importance of different endocytic pathways entered by a ligand. A full protocol to establish the isolated perfused rat liver using the system shown diagrammatically in *Figure 5* is given in *Protocol 8*.

Protocol 8. Establishment of the isolated perfused rat liver system

A. *Preparation of the system*

1. Wash human red blood cells twice with 0.1 M NaCl, 0.06 M $NaHCO_3$.
2. Mix 160 ml of packed red blood cells with 240 ml of perfusion medium[a] and place in the system (see *Figure 5*) to equilibrate for ~ 1 h.
3. Adjust the pump to deliver 1 ml/g expected liver weight/min to the perfusion reservoir and adjust the height of the perfusion reservoir above the liver position to deliver at this flow rate to the liver (baseline pressure).
4. Check that the temperature of the perfusate is 37 °C, the pH 7.4, the pCO_2 ~ 5 kPa, and the pO_2 ~ 12 kPa.

Protocol 8. *Continued*

B. *Operative procedure*

1. Use male Wistar rats weighing 150–200 g as donors.
2. Anaesthetize the donor animal with Hypnorm (fentanyl citrate 0.315 mg/ml and fluanisone; Janssen Pharmaceutical Ltd) 0.5 ml/kg intramuscularly.
3. Make 25-mm midline and transverse abdominal incisions in the anaesthetized rat, identify the bile duct, portal vein, and inferior vena cava after displacing the intestinal loops to the left side of the animal.
4. Maintain the operative field exposed with small retractors applied to the abdominal wall and adjacent viscera.
5. Occlude the distal end of the bile duct with a mosquito forceps and allow it to distend (~ 5 min).
6. Place loose ties around the proximal end of the portal vein and around the inferior vena cava, above the right renal vein.
7. Make a small incision in the wall of the bile duct (by now distended) at the distal end with a fine ophthalmic scissors.
8. Introduce a cannula in the bile duct (plastic tubing of 0.61 mm o.d., 0.20 mm i.d., 250 mm in length; Portex Ltd) with its tip placed 10 mm from the origin of the common bile duct; tie it with a 3/0 silk ligature and observe immediate free flow of bile.
9. Clamp the distal end of the portal vein with a mosquito forceps, make a small incision in the vein wall, introduce the perfusion cannula (plastic tubing 14 G by 280 mm with 16 G by 30 mm blunted end needle which has a groove at 6 mm from the distal end to facilitate ligation), tie it in place, and start perfusing immediately. The time from clamping the portal vein to the beginning of perfusion should last less than 15 sec so that tissue hypoxia is kept to a minimum.
10. Cut open the chest by a midsternal incision and cut off the thoracic vena cava above the diaphragm to drain the liver which must be uniformly perfused and not enlarged.
11. Tie off the inferior vena cava and cut off the vein below the ligature, effectively isolating the liver from the donor.
12. Free the liver from all attached tissues and transfer it with the bile duct cannula in place to its final position in the constant temperature cabinet, place it on its diaphragmatic side making sure that there is good drainage of perfusion medium, and avoid twisting the portal vein or the bile duct. Once in the perfusion cabinet cover the liver with a gauze soaked in normal saline (0.15 M NaCl) to avoid tissue dryness.
13. Observe uninterrupted bile production, adjust the perfusion pressure to

4 to 8 cm of height of perfusate above baseline pressure, start sodium taurocholate[b] infusion at 40 μmol/h and place the distal end of the bile cannula in a bile collecting vessel (1.5–ml plastic tube).

14. Allow 20 min for the liver to achieve 'steady state' before any experiments are conducted.

[a] Perfusion medium: DMEM containing 30 g/l BSA, 25 mM glucose, 50 000 U/l penicillin, 50 000 μg/l streptomycin, 1000 units/l heparin and 2.5 mM $CaCl_2$.
[b] Sodium taurocholate (Sigma), 2 mM in perfusion medium.

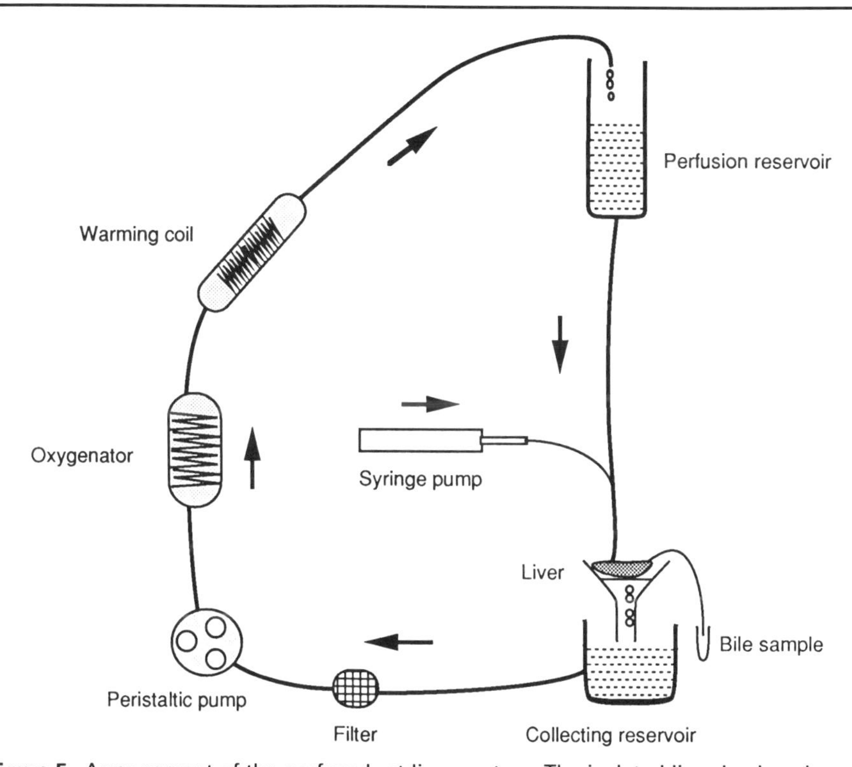

Figure 5. Arrangement of the perfused rat liver system. The isolated liver is placed on a fine plastic net above a plastic funnel which collects the perfusate leaving the liver into a collecting reservoir. From this a Watson–Marlow HRE 200 peristaltic pump (Watson Marlow Ltd) takes the perfusate via 12 G plastic tubing through an on-line blood filter, then through an oxygenator made of 80 feet of silastic medical grade tubing (Dow Corning cat. no. 602–235, Dow Corning Corp.) 0.058 inches (i.d.) by 0.077 inches (o.d.) arranged in six parallel lines and enclosed in a 240 × 80 mm plastic cylinder where it is equilibrated with 95% air: 5% CO_2 flowing at 0.1 l/min. On leaving the oxygenator, the perfusate passes through a heat exchanger set at 37 °C and is then pooled in a perfusion reservoir placed directly above the liver preparation and adjustable in height. The whole apparatus is maintained in an enclosed cabinet 800 × 800 × 600 mm with a constant ambient temperature of 37 °C provided by an electric fan heater coupled to a thermostat via a temperature probe. The syringe pump is a Treonic IP3 digital syringe pump (Vickers Ltd) connected to the perfusing catheter by a three-way tap.

The isolated liver should remain viable for a period of at least 3 h as judged by the following parameters.

- *Macroscopic appearance.* The liver colour remains uniform, of viable tissue appearance and non-oedematous.
- *Phenol red excretion.* Phenol red present in the perfusion medium at a concentration of 15 mg/l, appears in bile within 3 min of the start of the perfusion, and is continuously excreted throughout the experimental period.
- *Portal vein pressure.* This is estimated as the height of the perfusate column above the level observed when the perfusion medium is flowing freely, before being connected to the portal vein. Portal vein pressure is initially around 12 cm but within 15 min decreases to between 4 to 8 cm. A steady rise in portal vein pressure is seen in those livers with poor viability.
- *Oxygen consumption.* This is calculated from the difference of the pO_2 content in the perfusion medium before entering and after leaving the liver, the perfusate haemoglobin concentration, and the volume of perfusate given per min. Oxygen consumption should be approximately 3 μmol/min per gram of liver.
- *Bile flow.* The mean bile flow of the isolated liver is 1.4 ml/min per gram of liver.
- *Release of hepatocyte marker enzymes.* Alkaline phosphatase (ALP), alanine aminotransferase (ALT), and 5′nucleotidase (5′NT) may be measured in bile and perfusate (sampled from liver output) from isolated rat liver preparations (IPRL).

  Perfusion medium enzyme content should be: for ALP, 10–31 u/l (normal rat serum 246 u/l); for 5′NT, 4–9 u/l (normal rat serum 16 u/l); and for ALT, 8–91 u/l (normal rat serum 22 u/l). The moderate increase in ALT concentration may reflect progressive hepatocyte damage, while ALP is low when compared to normal rat as expected since the ALP from the perfusion medium lacks the bone component of the ALP present in normal young rats. The bile enzyme content in IPRL should be: for ALP, 17–22 u/l (normal rat bile 20 u/l); for 5′NT, 150–170 u/l (normal rat bile 150 u/l); and for ALT 27–28 u/l (normal rat bile 5 u/l).
- *Histological appearance.* After 3 h of perfusion, samples from perfused livers may be taken for histological examination and show preservation of the liver architecture and little cell damage.
- *Ligand transport from perfusate to bile.* When a single bolus injection of radiolabelled polymeric IgA is given to the isolated liver preparation, the ligand should be transported into bile with a time course similar to that seen in bile duct cannulated rats.
- *Marker enzyme distribution in subcellular fractions.* Subcellular fractionation on Ficoll and Nycodenz density gradients of livers perfused for 2 h

should show that distribution of *N*-acetyl-β-glucosaminidase, succinate dehydrogenase, and latent 5′NT is similar to that observed after subcellular fractionation of livers from freshly killed rats.

3.2.1 Subcellular fractionation to define endocytic compartments

The movement of ligands within cells at increasing times after a pulse dose can be followed by either electron microscopy of the tissue or by subcellular fractionation. Although both approaches are necessary for a complete understanding of endocytic pathways, fractionation also permits chemical examination of the various compartments and attempts at reconstitution of interactions between different organelles. In the authors' laboratory, we have used shallow, isosmotic, isopycnic gradients of Ficoll centrifuged in a vertical rotor to give a rapid separation of three different endocytic compartments, which are labelled sequentially by those ligands such as asialofetuin which are transferred to lysosomes for digestion (67, 68). The method is described in *Protocol 9*.

Protocol 9. Separation of endocytic compartments on a Ficoll gradient

A. *Preparation of gradients*

1. Dissolve Ficoll 400 (Pharmacia) by adding 1 ml of water/g at room temperature. Dialyse against a large volume of distilled water for 2 h and then adjust volume to give a 25% w/v solution. This may be frozen in aliquots and stored at −20 °C.
2. Place 4 ml of a solution of 45% Nycodenz[a] in the bottom of each Beckman 1 × 3.5 inch polyallomer Quick Seal centrifuge tube to act as a cushion.
3. Make linear gradients at 4 °C using 15 ml of each of 22% and 1% dialysed Ficoll solutions[b] for each tube. A simple gradient maker consists of two open chambers connected via a tap at the base and with an exit from the bottom of one chamber, which can be connected to a peristaltic pump. The chamber with the exit is stirred and becomes the mixing chamber. Put 15 ml of 22% Ficoll in the mixing chamber and 15 ml of 1% Ficoll in the other chamber. Open the tap between the chambers and simultaneously start the pump, which should run at about 1 ml/min. With a multi-channelled pump four gradients may be made simultaneously. The mixture is delivered to the centrifuge tubes through bent, large-bore hypodermic needles which rest in the centrifuge tube openings and drip freely. It is difficult and, in practice, unnecessary, to arrange for the gradient to run down the walls of vertical rotor tubes with their narrow openings. Leave the gradients at 4 °C overnight to smooth by diffusion.

Protocol 9. *Continued*

B. *Preparation of labelled post-mitochondrial supernatants*

1. Perfuse the liver thoroughly with ice-cold STM.[c] Remove to a beaker containing more STM to rinse the exterior and keep chilled.
2. Blot and weigh the liver. Transfer it to a beaker containing 3 ml/g liver of fresh cold STM and chop it with scissors.
3. Homogenize using a Potter Elvejhem homogenizer (Thomas Scientific) with a Teflon pestle rotating at 2400 r.p.m. and three complete up and down strokes. Keep the homogenizer vessel in iced water.
4. Centrifuge at 1500 *g* for 10 min at 4 °C to sediment nuclei and large mitochondria. Pour off the supernatant and keep chilled.

C. *Centrifugation*

1. Load 5 ml of post-mitochondrial supernatant to each gradient-containing tube using the peristaltic pump arrangement used to pump in the gradient at 4 °C.
2. Heat-seal the tubes and centrifuge them in a vertical rotor for an hour at 206 000 *g* at 4 °C using slow initial acceleration and no braking in the final stages of deceleration.

D. *Collection of fractions*

1. Cut the top off each tube with a scalpel.
2. Working in a 4 °C room, clamp the tube vertically and pass a 2-mm diameter stainless steel tube centrally and vertically from the top to the bottom of the tube. A simple plastic holder keeps the centrifuge tube vertical and allows the stainless steel tube to be clamped in place.
3. Suck the gradient out dense end first using a peristaltic pump running at about 1 ml/min and collect about 50 fractions using a suitable collector. Although this method of gradient collection is usually said to be poorer than upward displacement or tube piercing, we have found that it gives reproducible results as well as being quick and simple.
4. Check the refractive indices of small samples of sufficient fractions to gain an accurate picture of the exact density gradient in each tube. Reproducibility of peaks with respect to refractive index and hence density is excellent, although the precise fraction number at which a given density is found will vary from tube to tube.

[a] 45% Nycodenz: 45% Nycodenz (Nycomed) containing 10 mM *N*-tris (hydroxymethyl)-methyl-2-aminoethane sulphonic acid (TES) and 1 mM EDTA, pH 7.4 with NaOH.

[b] Ficoll solutions contain STE: 0.25 M sucrose, 10 mM TES, 1 mM EDTA, pH 7.4 with NaOH.

[c] STM: 0.25 M sucrose, 10 mM TES, 1 mM $MgCl_2$, pH 7.4 with NaOH.

Following a pulse dose of radiolabelled ligand, radioactivity appears first in sinusoidal plasma membrane (fractions of refractive index 1.3550–1.3600), but then moves sequentially to light endosomes, dense endosomes, and finally to the region on the cushion at the bottom of the gradient. The material on the cushion includes lysosomes, but can also be shown by recentrifugation on a 0–45% Nycodenz gradient to contain a third endosomal compartment, through which ligand flows before reaching true lysosomes. Radiolabel in this third endosomal compartment (very dense endosomes) appears at a much lower density on such gradients than the bulk of the lysosomal enzymes. In order to measure the amount of ligand in lysosomes at each time point, a sample of post-mitochondrial supernatant is therefore centrifuged on the Nycodenz gradient (*Protocol 10*) which clearly separates lysosomes from endosomes though it gives little separation of the three endosomal compartments from each other. The amount of label in very dense endosomes can then be calculated by subtracting the amount in true lysosomes from that in material of refractive index greater than 1.3730 on the Ficoll gradient. The passage of [^{125}I]asialofetuin through the endocytic compartments of rat hepatocytes analysed on the Ficoll and Nycodenz gradients is shown in *Figure 6*.

Protocol 10. Separation of lysosomes from endosomes on a Nycodenz gradient

1. Make gradients exactly as for Ficoll (see *Protocol 9*) placing 17 ml of the 45% Nycodenz solution in the mixing chamber and 17 ml STE in the other chamber for each gradient.
2. Load 2 ml of post-mitochondrial supernatant on to the gradient and continue as for Ficoll gradients.

Transcytosis of pIgA can also be followed using the shallow Ficoll and Nycodenz gradients. When subcellular fractionation is performed at different times after single-pass administration of [^{125}I]pIgA to isolated perfused rat livers, radioactivity is observed transiently in light endosomes and then in the denser compartments. If the experiments are performed on liver *in situ*, the light endosome fraction remains well labelled throughout the time course, because uptake of ligand by the liver does not approximate to the single-pass situation (in contrast to the uptake of [^{125}I]asialofetuin *in situ*), and the early (light) endosomes are continuously supplied with ligand from the circulation. In isolated perfused rat liver approximately 50% of the ^{125}I-pIgA taken up by the liver is transcytosed and 50% delivered to lysosomes. The transcytic pathway is particularly sensitive to microtubule disruption. Experiments in which microtubules were reversibly disrupted and sequential doses of

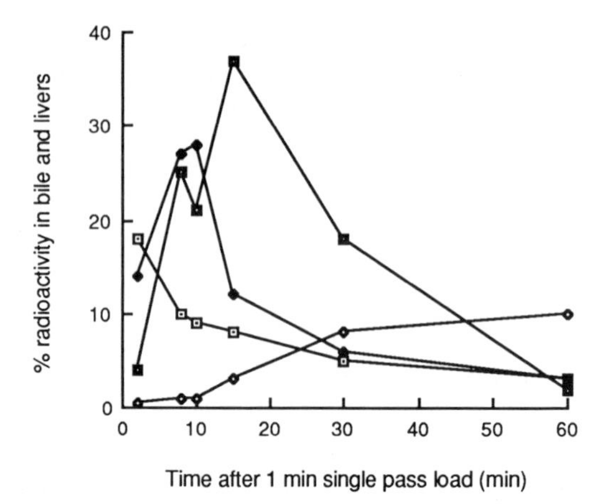

Figure 6. Time course of passage of ^{125}I-asialofetuin through endosome compartments in isolated perfused rat liver. □——□, light endosomes; ♦——♦, dense endosomes; ■——■, very dense endosomes; ◊——◊, lysosomes. Lysosomes rapidly digest asialofetuin to acid-soluble components which pass into bile and perfusate; hence the lysosome asialofetuin content only rises moderately.

differentially radiolabelled pIgA were given have shown that transcytosis does not require the ligand to enter the very dense endosomes and lysosomes (68).

3.2.2 Purification of endosomes

Enzyme assays showed that the major contaminant of both light and dense endosomal fractions taken from a Ficoll gradient was smooth endoplasmic reticulum membrane (marker enzyme glucose 6-phosphatase). Much of this could be removed by recentrifugation on Metrizamide gradients (*Protocol 11*). The composition of light and dense endosome fractions thus prepared is outlined in *Table 2*.

Protocol 11. Removal of smooth endoplasmic reticulum from endosomal fractions using Metrizamide

1. Place 11 ml of 12% Metrizamide[a] in a 1 × 3.5 inch polyallomer Quick Seal centrifuge tube. Using a syringe attached to a narrow steel tube long enough to each the base of the tube, underlay the first solution with 14 ml of 18% Metrizamide solution. Underlay the 18% Metrizamide with 4 ml of 45% Nycodenz solution (see *Protocol 9*).
2. Pool 5–6 fractions from the centre of the appropriate region of one or more Ficoll gradients (on the basis of refractive index measurements). Dilute with a solution of 10 mM TES, 1 mM EDTA, pH 7.4 until the refractive index is 1.354 or less.

3. Pump up to 10 ml of this diluted material on to the Metrizamide step gradient as described in *Protocol 9*. Material from three Ficoll gradients can usually be purified on two Metrizamide step gradients.
4. Seal tubes, centrifuge them for 1 h at 206 000 *g* and 4 °C and fractionate as described in *Protocol 9*.
5. Pool fractions containing most ligand (in the region of tubes 22–40), dilute threefold with STM (*Protocol 9*), and centrifuge for 60 min at 170 000 *g* to sediment endosomes.

[a] Metrizamide (centrifugation grade, Nycomed) solution contains 10 mM TES, 1 mM EDTA, pH 7.4 with NaOH.

Table 2. Composition of purified endosome fractions

Fraction	Refractive index band on Ficoll gradient	Yield (% homogenate)			
		^{125}I ligand in appropriate homogenate	Protein	Glucose 6-phosphatase	Lysosomal enzymes
Light endosomes	1.3600–1.3660	25.0	0.7	0.4	0.5
Dense endosomes	1.3661–1.3730	26.0	0.5	4.0	0.7

Very dense endosomes cannot easily be purified from a Ficoll gradient since they are mixed with all the other material that passes through 22% Ficoll to reach the Nycodenz cushion. These vesicles are also much more fragile when sedimented into a pellet than light or dense endosomes, leaking both labelled ligand and enzyme. They are best prepared on a step gradient as described in *Protocol 12*.

Protocol 12. Preparation of very dense endosomes

1. Dilute some 45% Nycodenz (see *Protocol 9*) with STE to a Nycodenz concentration of 20%.
2. Place 14 ml of dialysed 20% Ficoll solution (see *Protocol 9*) in a 1 × 3.5 inch polyallomer Quick Seal centrifuge tube. Underlay as described in *Protocol 11* with 14 ml of the 20% Nycodenz solution and finally with 4 ml of 45% Nycodenz solution.
3. Pump on 7 ml of post-mitochondrial supernatant and centrifuge and fractionate as described in *Protocol 9*.

The very dense endosomal fraction bands between the 20% Ficoll and the 20% Nycodenz layers. The other endosomal fractions do not penetrate the 20% Ficoll layer whilst lysosomes pass through the 20% Nycodenz. Contaminating endoplasmic reticulum can be removed by recentrifugation on a Metrizamide step gradient as described for light and dense endosomes.

3.2.3 Characterization of endosomes

No endosomal fraction has yet been shown to posses a characteristic marker enzyme or indeed protein of any kind. They are therefore identified primarily by their ligand content. However, they possess some enzyme activities and receptor proteins, which are also found elsewhere in the cell. Thus light endosomes possess latent 5′-nucleotidase and a [Mg^{2+}]ATPase that resembles the protein-pumping ATPase of chromaffin granules as well as the receptors for polymeric IgA and asialoglycoprotein (69). All three endosome fractions contain mannose 6-phosphate receptors. Dense endosomes have some proteolytic enzymes but very little *N*-acetyl-*β*-glucosaminidase; very dense endosomes have more *N*-acetyl-*β*-glucosaminidase although still very much less than lysosomes.

Further characterization of the endocytic compartments can be gained by examining their properties in cell-free systems. The authors have been using such a system to examine the interaction with lysosomes of endosomes loaded *in vivo* with ^{125}I-labelled ligand. After reaction *in vitro*, lysosomes and endosomes are separated on small Nycodenz gradients (*Protocol 13*) and the radioactivity associated with each organelle is measured.

Protocol 13. Analysis of endosome–lysosome interaction by separation on small Nycodenz gradients

1. Mix 45% Nycodenz and STE (see *Protocol 9*) to give a solution containing 35% Nycodenz.
2. Make linear gradients at 4 °C in Sorvall 6-ml polyallomer Ultracrimp tubes using 2.4 ml of 35% Nycodenz and 2.4 ml STE solutions for each gradient. Place the 35% Nycodenz in the mixing chamber and proceed as for Ficoll gradients (see *Protocol 9*).
3. Run the *in vitro* reaction in a volume of 0.5 ml. Chill to stop.
4. Load the reaction mixture in the cold, crimp seal the tubes, and centrifuge for 20 min at 370 000 *g* and 4 °C.
5. Cut the tops from the tubes and collect five drop fractions using a 1 mm diameter stainless steel tube in a small version of the apparatus described for gradient collection in *Protocol 9*.
6. Count the fractions, measure refractive indices to check extact shapes of gradients, and, if desired, measure *N*-acetyl-*β*-glucosaminidase activities to check positions of lysosomes.

The first experiments were done using post-mitochondrial supernatants as a source of both endosomes and lysosomes (70). More recently (66), it has been shown with purified fractions that dense endosomes will react directly with purified lysosomes (71) when incubated at 37 °C. For maximum transfer of label, cytosol and an ATP-regenerating system are required.

Whilst the centrifugal assay indicates association of endosomes with lysosomes, it does not demonstrate that fusion occurs between the two compartments. In order to examine the potential abilities of membranes from different fractions to fuse with each other *in vitro* the authors have been using fluorescent dequenching of octadecyl-rhodamine B-chloride (R18; reference 72). Purified endosomes are loaded with self-quenching concentrations of R18; fusion with another unlabelled membrane fraction results in dilution of the R18 and a consequent relief of self-quenching (*Protocol 14*).

Protocol 14. Measurement of membrane fusion by relief of octadecyl-rhodamine B-chloride (R18) self-quenching

1. Prepare a 20 mM solution of R18 (Molecular Probes Inc, Cambridge Bioscience) in ethanol at room temperature or just above.
2. Add 3 μl to 1 ml of a suspension of endosomes (0.5–1 mg protein) in STM (see *Protocol 9*) which is also at room temperature or above and mix very rapidly and thoroughly.
3. Leave at room temperature for 15 min with gentle mixing, protected from light.
4. Apply to a small column of Sepharose CL-4B approximately 6 cm high and equilibrated with STM. This can be made in a Pasteur pipette.
5. Wash with the STM and collect the coloured eluate. Free R18 remains bound to the Sepharose. The R18-loaded membranes are collected in a volume of 1.2–1.3 ml.
6. Prepare incubation mixtures containing receptor membranes, ATP, inhibitors, etc. in a final volume of 0.9 ml of STM. Incubate these and the R18-loaded membranes at 37 °C for 2 min; then start the reactions by adding 0.1 ml of R18-loaded membrane suspension to each incubation mixture and mixing thoroughly.
7. Read the fluorescent emissions at 590 nm using an excitation wavelength of 560 nm at increasing time intervals. Protect from direct light.
8. At the end of the experiment add 60 μl 20% Triton X-100 to each tube; mix and read the emission again to give a measure of total fluorescence in that tube.
9. Calculate the emissions as a percentage of the total emission (allowing for the volume increase on addition of Triton). Correct for the small leakage

Protocol 14. *Continued*

or redistribution of R18 (approximately 1–2%/h) by including tubes which contain all components except the acceptor membranes.

Using this method, the authors have shown, for example, that incubation of R18-labelled dense endosomes with lysosomes and ATP at 37 °C leads to 25–30% dequenching, whilst incubation with a similar protein concentration of a crude mitochondrial preparation produces only about 5% dequenching. Incubation at 4 °C produces no dequenching.

4. Conclusions

There is now a wealth of techniques to study membrane traffic pathways in polarised epithelial cells. At the biochemical level, the availability of reliable subcellular fractionation methods, domain-specific cell-surface labelling, and good electrophoretic analysis has already led to the identification of specific proteins moving along particular membrane traffic pathways. The molecular cloning of such proteins followed by molecular cut-and-paste experiments will lead to the identification of targeting signals required for movement along specific routes and/or to specific destinations. In addition, the further developmnt of cell-free systems (73) and permeabilized cell systems (74, 75), which allow manipulation of the cytoplasmic compartment while leaving overall morphology and intracellular organization intact, will establish the nature of membrane and cytosolic proteins required for vesicular membrane traffic.

Acknowledgements

The authors' experimental work quoted in this chapter has been funded variously by the Medical Research Council, the Cancer Research Campaign, Ciba-Geigy Pharmaceuticals, and the East Anglian Regional Health Authority.

References

1. Berridge, M. J. and Oschmann, J. L. (1972). *Transporting epithelia*, p. 91. Academic Press, New York.
2. Simons, K. and Fuller, S. D. (1985). *Ann. Rev. Cell Biol.* **1**, 243.
3. Rodriguez-Boulan, E. and Nelson, W. J. (1989). *Science* **245**, 718.
4. Hauri, H.-P., Sterchi, E. E., Bienz, D., Fransen, J. A. M., and Marxer, A. (1985). *J. Cell Biol.* **101**, 838.
5. Bartles, J. R., Feracci, H. M., Stieger, B., and Hubbard, A. L. (1987). *J. Cell Biol.* **105**, 1241.

6. Fuller, S., von Bonsdorff, C. H., and Simons, K. (1984). *Cell* **38**, 65–77.
7. Wandinger-Ness, A., Bennett, M. K., Antony, C., and Simons, K. (1990). *J. Cell Biol.* **111**, 987.
8. Caplan, M. and Matlin, K. S. (1989). *Mol. Cell Biol.* **8**, 71.
9. Matter, K., Brauchbar, M., Bucher, K., and Hauri, H.-P. (1990). *Cell* **60**, 429.
10. Le Bivic, A., Quaroni, A., Nichols, B., and Rodriguez-Boulan, E. (1990). *J. Cell Biol.* **111**, 1351.
11. Simons, K. and Wandinger-Ness, A. (1990). *Cell.* **62**, 207.
12. Evans, W. H., Flint, N. A., and Vischer, P. (1980). *Biochem. J.* **192**, 903.
13. Hubbard, A. L., Stieger, B., and Bartles, J. R. (1989). *Ann. Rev. Physiol.* **51**, 755.
14. Hidalgo, I. J., Raub, T. J., and Borchardt, R. T. (1989). *Gastroenterology* **96**, 736.
15. de Duve, C. (1971). *J. Cell Biol.* **50**, 20D.
16. Moktari, S., Feracci, H., Gorvel, J-P., Mishal, Z., Rigal, A., and Maroux, S. (1986). *J. Membrane Biol.* **89**, 53.
17. Stieger, B., Marxer, A., and Hauri, H.-P. (1986). *J. Membrane Biol.* **91**, 19.
18. Howell, K. E., Devaney, E., and Gruenberg, J. (1989). *Trends biochem. Sci.* **14**, 44.
19. Stieger, B. and Murer, H. (1983). *Eur. J. Biochem.* **135**, 95.
20. Colas, B. and Maroux, S. (1980). *Biochim. Biophys. Acta* **600**, 406.
21. Dahlquist, A. (1968). *Anal. Biochem.* **22**, 99.
22. Maguire, G. A., Docherty, K., and Hales, C. N. (1983). *Biochem. J.* **212**, 211.
23. Peters, T. J. (1976). *Clin. Sci. mol. Med.* **51**, 557.
24. Bretz, R., Bretz, H., and Palade, G. E. (1980). *J. Cell Biol.* **84**, 87.
25. Pennington, R. J. (1961). *Biochem. J.* **80**, 649.
26. Laemmli, U. K. (1970). *Nature* **227**, 680.
27. Sargiacomo, H., Lisanti, M., Graeve, L., Le Bivic, A., and Rodriguez-Boulan, E. (1989). *J. Membrane Biol.* **107**, 277.
28. Findlay, J. B. C. (1987). In *Biological membranes: a practical approach* (ed. J. B. C. Findlay and W. H. Evans), pp. 197–211. IRL Press, Oxford.
29. Sullivan, H. K. and Williams, P. R. (1982). *Anal. Biochem.* **120**, 254.
30. Richardson, K. and Parker, C. D. (1985). *Infection Immunity* 87.
31. Morrison, M. (1974). In *Methods in enzymology*, Vol. 32B (ed. Fleischer, S. and Padzer, L.), p. 103–109. Academic Press, New York and London.
32. Stoorvogel, W., Geuze, J. H., Griffith, M. J., Schwartz, L. A., and Strous, J. G. (1989). *J. Cell Biol.* **108**, 2137.
33. Grisolia, S., Hernandez-Yago, J., and Knecht, E. (1988). *Curr. Topics Cell Regulation* **27**, 387.
34. Lisanti, P. M., Le Bivic, A., Sargiacomo, M., and Rodriguez-Boulan, E. (1989). *J. Cell. Biol.* **109**, 2117.
35. Le Bivic, A., Real, X. F., and Rodriguez-Boulan, E. (1989). *Proc. nat. Acad. Sci., USA* **86**, 9313.
36. Le Bivic, A., Sambuy, Y., Mostov, K., and Rodriguez-Boulan, E. (1990). *J. Cell Biol.* **140**, 1533.
37. Busch, G., Hoder, D., Reutter, W., and Tauber, R. (1989). *Eur. J. Cell Biol.* **50**, 257.
38. Matter, K., Brauchbar, M., Bucher, K. and Hauri, H. P. (1990). *Cell* **60**, 429.

39. Hurley, W. L., Finkelostein, E., and Holst, B. D. (1985). *J. immunol. Methods* **85**, 195.
40. Bordier, C. (1981). *J. Biol. Chem.* **256**, 1604.
41. Pryde, J. G. (1986). *Trends biochem. Sci.* **11**, 160.
42. Towbin, H., Staehelin, T., and Gordon, J. (1979). *Proc. nat. Acad. Sci. USA* **76**. 4350.
43. Roth, R., Maddux, B. A., Cassell, D. J., and Goldfine, I. D. (1983). *J. Biol. Chem.* **258**, 12094.
44. Salas, I. J. P., Misek, E. D., Vega-Salas, E. D., Gunderson, D., Gereijido, M., and Rodriguez-Boulan, E. (1986). *J. Cell Biol.* **102**, 1853.
45. Abraham, G. and Low, P. S. (1980). *Biochim. Biophys. Acta* **597**, 285.
46. Gahmberg, G. and Andersson, C. L. (1977). *J. Biol. Chem.* **252**, 5888.
47. Passanti, A. and Hart, G. W. (1988). *J. Biol. Chem.* **263**, 7591.
48. Brandli, W. A., Hansson, C. G., Rodriguez-Boulan, E., and Simons, K. (1988). *J. Biol. Chem.* **263**, 16283.
49. Hjelmeland, M. L., Nebert, W. D., and Chrambach, A. (1979). *Anal. Biochem.* **95**, 201.
50. Ames, G. F. L. and Nikaids, K. (1976). *Biochemistry* **15**, 616.
51. Perdew, G. H., Schaup, H. W., and Selivonchick, D. P. (1983). *Anal. Biochem.* **135**, 453.
52. O'Farrell, P. H. (1975). *J. Biol. Chem.* **250**, 4007.
53. Bravo, R. (1984). *Proc. nat. Acad. Sci., USA* **81**, 4848.
54. Brandli, A. W. and Simons, K. (1989). *EMBO J.* **8**, 3207.
55. Hughson, J. E. and Hopkins, C. R. (1990). *J. Cell Biol.* **110**, 337.
56. Dix, C. J., Obray, H. Y., Hassan, I. F., and Wilson, G. (1987). *Biochem. Soc. Trans.* **15**, 439.
57. Bomsel, M., Prydz, K., Parton, R. G., Gruenberg, J., and Simons, K. (1989). *J. Cell Biol.* **109**, 3243.
58. Brandli, A. W., Parton, R. G., and Simons, K. (1991). *J. Cell Biol.* **111**, 2909.
59. Hunziker, W. and Hellman, I. (1989). *J. Cell Biol.* **109**, 3291.
60. Benford, D. J. and Hubbard, S. A. (1987). In *Biochemical toxicology: a practical approach* (ed. K. Snell and B. Mullock), pp. 65–8. IRL Press, Oxford.
61. Maurer, H. R. (1986). In *Animal cell culture: a practical approach* (ed. R. I. Freshney), p. 23. IRL Press, Oxford.
62. Wada, I., Himeno, M., Furuno, K., and Kato, K. (1986). *J. Biol. Chem.* **261**, 2222.
63. Baron, M. D. and Luzio, J. P. (1987). *Biochim. Biophys. Acta* **927**, 81.
64. Geuze, H. J., van der Donk, H. A., Simmons, C. F., Slot, J. W., Strous, G. J., and Schwartz, A. L. (1986). *Int. Rev. exp. Pathol.* **29**, 113.
65. Sztul, E., Kaplin, A., Saucan, L., and Palade, G. (1991). *Cell* **64**, 81.
66. Luzio, J. P. and Mullock, B. M. (1991). In *Proceedings of the second European Workshop on Endocytosis* (ed. P. Courtoy). Springer Verlag, Berlin. (In press.)
67. Branch, W. J., Mullock, B. M., and Luzio, J. P. (1987). *Biochem. J.* **244**, 311.
68. Perez, J. H., Branch, W. J., Smith, L., Mullock, B. M., and Luzio, J. P. (1988). *Biochem. J.* **251**, 763.
69. Mullock, B. M., Hinton, R. H., Peppard, J. V., Slot, J. W., and Luzio, J. P. (1987). *Cell Biochem. Function* **5**, 235.

70. Mullock, B. M., Branch, W. J., van Schaik, M., Gilbert, L. K., and Luzio, J. P. (1989). *J. Cell Biol.* **108**, 2093.
71. Maguire, G. A. and Luzio, J. P. (1985). *FEBS Lett.* **190**, 122.
72. Hoekstra, D. (1990). *Hepatology* **12**, 615.
73. Gruenberg, J. and Howell, K. E. (1989). *Ann. Rev. Cell Biol.* **5**, 453.
74. Becker, C. J. H. and Balch, W. E. (1989). *J. Cell Biol.* **108**, 1245.
75. Simons, K. and Virta, H. (1987). *EMBO J.* **6**, 2241.

3

Sorting between exocytic pathways in PC12 cells

LOUISE P. CRAMER and DANIEL F. CUTLER

1. Introduction

Exocytic protein traffic in neuronal and endocrine cells is complicated by the presence of more than one route to the cell surface. In addition to the constitutive secretory pathway, regulated secretory routes are also present in these cells. Regulated secretion is characterized by the intracellular storage of secretory material which is only released from the cell following the application of an external stimulus. This is in contrast to constitutive secretion in which proteins are not stored inside the cell but are constantly released. The organelles acting as storage compartments in neuronal and endocrine cells are both striking and characteristic of the cell type. In endocrine cells, dense core granules (DCGs) store secretory proteins, peptides, and small molecules such as classical neurotransmitters. In neurons, synaptic vesicles store neurotransmitters but not proteins.

The coexistence of both regulated and constitutive secretory pathways in one cell was demonstrated by Kelly and co-workers in a DCG-containing cell line (AtT20) derived from the anterior pituitary (1, 2). This demonstration implied that regulated secretory cells must be able to sort regulated secretory proteins and constitutive secretory proteins so that only regulated secretory proteins are targeted to DCGs during granule formation. Later the coexistence of both pathways was also demonstrated in other regulated secretory cells.

Sorting in neuronal cells is further complicated because these cells have some DCGs in addition to synaptic vesicles (3). Endocrine cells also contain small clear vesicles which are like synaptic vesicles (synaptic-like vesicles; SLVs) in terms of morphology, physical properties, and membrane protein composition. These observations suggest that neuronal and endocrine cells may have to sort proteins between three exocytic pathways (*Figure 1*). The relationships between the different exocytic organelles in terms of membrane traffic are not yet clear.

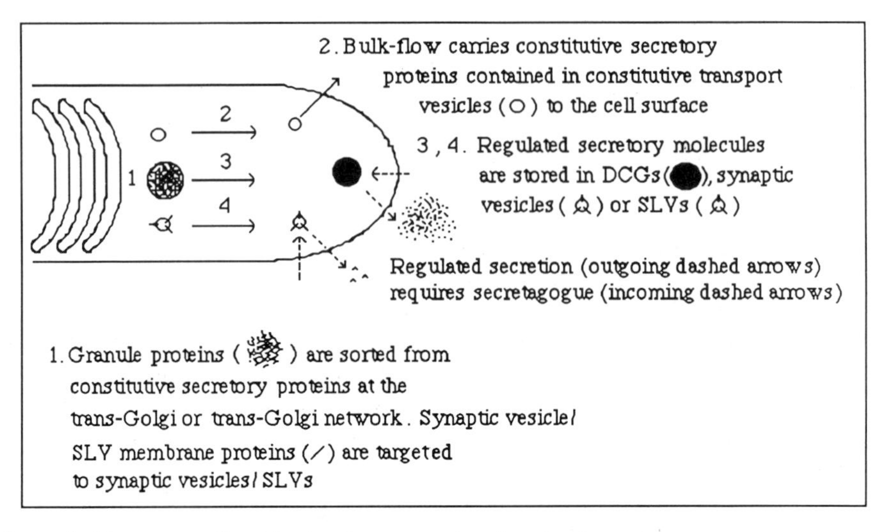

Figure 1. Three exocytic pathways in neuronal and endocrine cells.

2. Systems available

To investigate sorting between exocytic pathways in endocrine or neuronal cells it is advantageous to study targeting to DCGs, SLVs, or synaptic vesicles in a single cell type that can be manipulated biochemically and morphologically with relative ease. Isolation of a homogeneous population of endocrine or neuronal cells from tissues is difficult, and a mixed population of cells with a finite life in culture makes many of the experiments described below very hard to perform. However, such experimentation is possible in endocrine- or neuronal-like cells that are continuously growing in culture as a homogeneous population (i.e. a clonal cell line), common examples of which are at AtT20 cells (derived from mouse corticotrophs, producing proopiomelanocortin), GH3 cells (derived from rat somatotrophs, producing growth hormone), and PC12 cells (see below). Other less common examples are RIN5F cells and Neuro2A cells. As mentioned in the introduction, targeting to regulated secretory organelles has been studied extensively in AtT20 and PC12 cells. In addition, AtT20 and GH3 cells have been used to study proteolytic processing of peptides and proteins stored in DCGs. PC12 cells do not process prohormones efficiently. One requirement for the study of targeting to regulated secretory organelles is to be able to isolate these organelles. PC12 cells have more DCGs than GH3 cells and granules from PC12 cells may be easier to isolate than those of AtT20 cells. This chapter will describe the manipulation of PC12 cells for the investigation of sorting between exocytic pathways.

PC12 cells are a clonal cell line derived from a transplantable tumour of rat adrenal medulla (5). These cells can exist in two states, undifferentiated and differentiated. Undifferentiated cells are endocrine-like cells and differentiated cells more like neurons. PC12 cells are therefore a useful tissue culture mammalian model for the study of targeting to regulated secretory organelles in endocrine and neuronal cells.

Undifferentiated PC12 cells are rather round in appearance with heterogeneous morphology and often grow in clumps. This is illustrated in *Figure 2a*. They behave like a population of chromaffin cells of the adrenal medulla (6). DCGs in chromaffin cells are approximately 350 nm in diameter. Chromaffin DCGs store catecholamines (dopamine, noradrenaline, adrenaline), secretory proteins (for example, the secretogranins or chromogranins, some catecholamine-synthesizing enzymes, and low levels of small peptides such as endorphin, enkephalin, neuropeptide Y, neurotensin, and somatostatin), nucleotides, and calcium (6–8). DCGs in PC12 cells are more varied in diameter. They also store catecholamines (but not adrenaline), secretory proteins, small peptides (but less than chromaffin cells), nucleotides, and calcium. It is thought that PC12 cells represent immature chromaffin cells in varying stages of maturation. Stored products can be released from

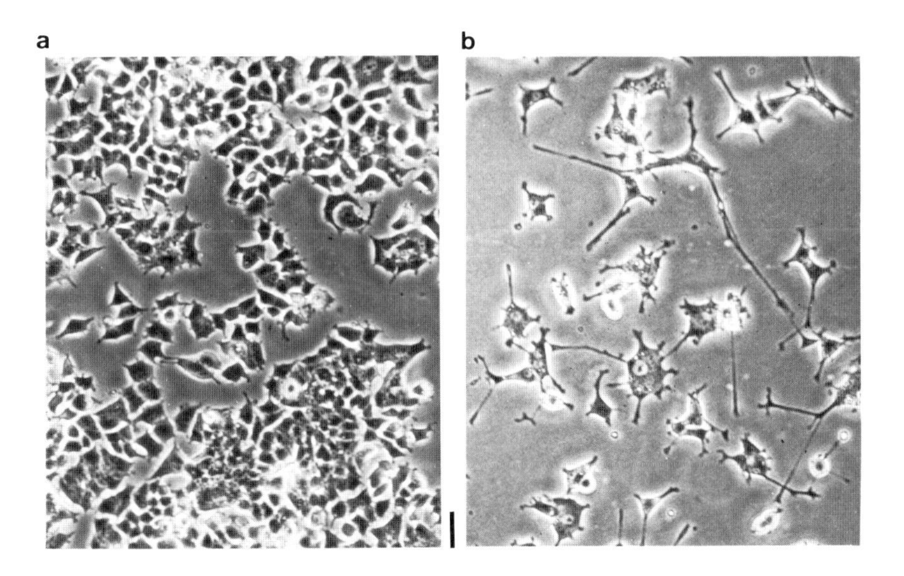

Figure 2. Light microscopy of PC12 cells. PC12 cells growing on plastic in tissue culture medium in (a) the absence of (b) the presence of NGF (treatment for 3 days) were viewed under phase optics and photographed with the dish lid removed. Scale bar, 50 μm. Note that, in the absence of NGF, cells can grow singly (and are flatter) or in clumps (and are rounder). Cells rarely cover the dish as a uniform monolayer. (a) represents the range of densities typically obtained on a dish as cells approach confluency. (b) In the presence of NGF, cells flatten considerably and project neurite-like processes from the cell body.

chromaffin and PC12 cells by, for example, depolarizing potassium or cholinergic stimulation (6, 9, 10).

PC12 cells differentiate into cells resembling sympathetic-like neurons in the presence of nerve growth factor (NGF) (5, 11, 12). This is illustrated in *Figure 2b*. The process does not seem to require DNA replication and is reversible. PC12 cells extend neurite-like processes away from the cell body at a rate of 20–40 μm a day and stop replicating, although not immediately. Neurite extension seems to depend on protein kinase A and protein kinase C dependent pathways (13). DCGs are maintained while SLVs can be partially induced by NGF (5, 6, 11, 12). SLVs can be observed in clusters, often in neurite-like process endings or varicosities. Differentiated cells also respond to depolarizing potassium or cholinergic stimulation (6, 9, 10). This chapter will mostly concentrate on methods developed in undifferentiated PC12 cells although they can also be applied to differentiated PC12 cells.

3. Culturing PC12 cells

3.1 General techniques

PC12 cells are sensitive and should be handled with care. It should also be borne in mind that there are several PC12 subclones and so this sensitivity may vary. The following culturing procedures have been developed for PC12 cells obtained from the Cell Culture Facility, University of California, San Francisco, but should generally apply to all PC12 cells. Routine PC12 tissue culture methods are illustrated in *Protocol 1*.

Protocol 1. Routine PC12 tissue culture

Materials

- PC12 cells obtained from the Cell Culture Facility, University of California, San Francisco
- 'Falcon' (Becton Dickinson Labware), 'Costar' (Costar), or 'Nunc' (Gibco) plastic ware
- Dulbecco's modified minimum Eagle's medium (DMEM) or Roswell Park Memorial Institute (RPMI) 1640 medium supplemented with 10% horse serum which has been previously heat inactivated at 55 °C for 30 min and 5% fetal calf serum (FCS). Medium supplemented with sera will be called culture medium in this article.
- trypsin/EDTA solution (0.25% trypsin, 0.02% EDTA in phosphate-buffered saline (PBS) pH 7.4)
- freezing mix: 90% FCS/10% dimethyl sulphoxide (DMSO)
- equipment: tissue culture incubator set at 10% carbon dioxide; 37 °C water bath; inverted microscope; low-speed benchtop centrifuge

Method

1. To passage cells[a], remove culture medium by aspiration, rinse cells very briefly using 5 ml warm trypsin/EDTA solution, and then detach them from the dish by incubation at 37 °C with 1 ml fresh trypsin/EDTA solution per 10-cm dish. This takes 2–10 min in the incubator depending on the state of the cells and the age of the trypsin. The trypsin should then be diluted about fivefold by the addition of fresh culture medium to the dish and the cells removed and clumps broken up by trituration.[b] The cell suspension is then pipetted into a 15-ml sterile centrifuge tube, the cells collected by centrifugation at 1000 *g* for 5 min, the supernatant removed, the cells resuspended in fresh warm culture medium, and plated out.
2. Cells to be used in 2–4 days time should be plated at 3×10^6 cells per 10-cm dish, in 7 ml of medium and fed every 2–3 days. *Figure 2a* shows cells that are suitable for use 3 days after plating at this density. Growth rate depends on sera batches, particular culture medium used and source of plastic ware.
3. To store cells, instead of resuspending in fresh culture medium after centrifugation, resuspend in 90% FCS/10% DMSO (ice-cold, 1 ml/10-cm dish) and transfer to a sterile freezing vial. Wrap the vial in an insulating material such as cottonwool, place in an insulating box, e.g. polystyrene, leave at −70 °C for 3 days to 1 month, and then transfer to liquid nitrogen.
4. To restart culture from frozen stocks, rapidly thaw frozen cells by immersion in a water bath set at 37 °C, then dilute into 10 ml culture medium. The cells are then collected by centrifugation and plated in fresh culture medium (as above) usually at one vial per 10-cm dish. It should be possible to experiment on cells frozen and thawed in this way 6–12 days after thawing.
5. For some experiments cells can be grown in the presence of NGF to cause differentiation. When required, NGF (either the 2.5S or 7S forms may be used) should be added to a final concentration of about 50 ng/ml. Morphological changes or changes in release of stored products or gene expression induction are usually observable after about a day of treatment but the effect may increase after longer treatments. However, the effects may also decrease after maximum induction has been reached if treatment is continued.

[a] PC12 cells tend to be most healthy if they are not 100% confluent when passaged and are not plated at very low densities. Passaging cells (even those at low density) if they are particularly unhealthy, i.e. many are detaching from the dish, can improve their viability.

[b] PC12 cells tend to grow as clumps (see *Figure 2a*) which will persist during trypsinization, but cells should not be forced completely into a single-cell monolayer. This is because there is evidence that a high cell density and an increased length of time after plating contributes to a high level of inducible regulated secretory activity. Note, however, that cells will detach from the dish

Protocol 1. *Continued*

if allowed to become too clumpy. To increase separation, following removal of the trypsin, pass the cells up and down in the fresh culture medium (7 ml) 5–10 times in a 10-ml glass pipette applying gentle pressure on the bottom of the centrifuge tube before plating. Too little pressure does not separate cells efficiently. Too much pressure is damaging to cells.

3.2 Growing cells for labelling and morphological experiments

Growing cells for labelling experiments is illustrated in *Protocol 2* and for fluorescence experiments in *Protocol 3*. To take phase photographs of these cells, an ordinary single lens reflex camera can be attached to an inverted microscope.

Protocol 2. Growing cells for labelling experiments

Materials

- general culture reagents (listed in *Protocol 1*)
- methionine/cysteine-free minimum essential medium (MEM)
- sulphate-free MEM
- radioactive labels, e.g. [^{35}S]methionine/[^{35}S]-cysteine, [^{35}S]sulphate, [^{3}H]-dopamine, or [^{3}H]noradrenaline

Method

1. To label DCGs with catecholamines[a], rinse the dishes of cells twice in fresh culture medium then incubate with [^{3}H]dopamine or [^{3}H]noradrenaline in culture medium for 1–2 h. This is the simplest and most convenient method for marking DCGs.
2. To label dishes of cells long-term (3–24 h) with radiolabelled amino acids such as [^{35}S]methionine or [^{35}S]cysteine, incubate cells with label in methionine-free minimum essential medium (MEM) or methionine or cysteine-free medium made from a kit, e.g. RPMI-1640 SelectAmine Kit (Gibco UK) supplemented with between 1 and 10% culture medium. The level of supplementation will vary depending on the length of incubation, e.g. 5% for an overnight labelling.
3. To pulse-label cells, first rinse the cells 2–3 times with the MEM or RPMI medium, then starve them of labelled amino acid for 15–30 min in the MEM or RPMI medium, and then label as in step **3**; except do not supplement this medium with culture medium.[b]
4. To label cells with $^{35}SO_4$, sulphate-free medium with reduced levels of methionine and cysteine is required. To generate this replace the $MgSO_4$

in methionine and cysteine-free medium with $MgCl_2$ and then add methionine and cysteine to 1% of normal levels (for pulse labelling) or add 1% culture medium (for long-term labelling).

(a) Wash cells with the appropriate sulphate-free medium and then incubate in this medium for 15–30 min.

(b) For a pulse label, add 800 μCi–1 mCi/ml [^{35}S]sulphate for 5–10 min.

(c) For a long-term label, add 200–500 μCi/ml for 3–24 h.

[a] Catecholamines labelled on the aromatic ring tend to be more stable in storage (stored in aliquots at −70 °C).

[b] In these experiments, a 3-cm dish is labelled in a volume of 1 ml with 1 μCi catecholamine, or 100–250 μCi [^{35}S]amino acid; a 6-cm dish is labelled in a volume of 2 ml with 3 μCi catecholamine, or 250–500 μCi [^{35}S]amino acid; a 10-cm dish is labelled in a volume of 5 ml with 5 μCi catecholamine, or 500 μCi-1 mCi[^{35}S]amino acid.These levels of label are sufficient to allow rapid visualization of relatively rare membrane proteins following immunoprecipitation. In general, use the lower level of radioactive amino acid for cells to be labelled long-term.

Protocol 3. Growing cells for immunofluorescence

Materials

- general culture reagents (listed in *Protocol 1*)
- sterile glass coverslips
- sterile poly-L-lysine (0.5 mg/ml in water)
- NGF

Method

1. Place sterile glass coverslips in culture dishes.
2. Treat coverslips with sterile 0.5 mg/ml poly-L-lysine[a] (it is sufficient to reconstitute the solid in sterile water and to store in a sterile container) for 15–30 min at 37 °C, remove, and then briefly rinse coverslips on the dish once in culture medium.[b]
3. Plate cells on to coverslips as in *Protocol 1*. Aim to use cells within 2–5 days of plating because cells start to look unhealthy after extended periods of time on poly-L-lysine.[c]

[a] Poly-L-lysine causes cells to flatten somewhat. It can be reused 2–3 times.

[b] Cover three-quarters of the coverslips with poly-L-lysine by adding a drop to the centre of each coverslip from a sterile Pasteur pipette. This prevents poly-L-lysine covering the dish, thus encouraging cells to preferentially grow on the coverslips.

[c] As shown in *Figure 2b* NGF-treated cells are flatter than untreated cells. Immunofluorescence is therefore generally easier on cells that have been treated with NGF.

If cells start to detach from the dish or coverslip during any procedure, reduce the 'rinsing' or 'washing' steps. Also remove cells found in any assay solution by centrifugation (about 10 sec in a microcentrifuge at 14 000 r.p.m. or 2–5 min in a bench-top centrifuge at 1000 r.p.m.) and, if necessary, add to any cells to be lysed in Nonidet P40 or NDET (e.g. *Protocol 4*).

4. Stimulation of secretion

To study sorting between exocytic pathways it is first necessary to demonstrate that the cells cultured in the laboratory have both regulated and constitutive secretory pathways (*Protocols 4* and *5*). Some regulated secretory markers stored in PC12 DCGs are: catecholamines (6); chromogranins/secretogranins (15); transfected growth hormone (16); transfected prosomatostatin (17), and transfected insulin (see reference 18). Some PC12 constitutive secretory markers are: laminin (16); a heparin sulphate proteoglycan (19); and transfected kappa light chain (18).

Several reagents can be used to stimulate calcium-dependent secretion from PC12 cells. Common ones are carbamylcholine (carbachol) which acts through the acetylcholine receptor, and high potassium (*Protocol 4*). Other secretagogues such as barium in the absence of calcium (20) can also be used.

Protocol 4. Inducing regulated secretion from cells labelled with [^{3}H]catecholamines

Materials

- general culture reagents (listed in *Protocol 1*)
- radiolabelled cells (see *Protocol 2*, step **1**)
- culture medium supplemented with 5 mM carbachol or 55 mM potassium chloride
- 1% Nonidet P40 (NP40) in PBS or NDET: 1% NP40; 0.6% deoxycholate; 66 mM EDTA; 10 mM Tris, pH 7.5. When making NDET, dissolve EDTA first and bring the pH back to 7.4 before adding the other ingredients
- aqueous scintillant; scintillation vials/inserts; scintillation counter

Method

1. Wash labelled cells for 10–20 min in several changes of culture medium. If cells start to detach from the dish reduce the washing time.
2. Add 1–4 ml culture medium (depending on the size of the dish) containing 5 mM carbachol or 55 mM KCl kept at 37 °C (stimulating medium). DMEM itself contains 1.8 mM $CaCl_2$.
3. Replace stimulating medium every 2 min for 12–18 min, retaining each

change of medium on ice for subsequent analysis. Dishes can be kept on the bench at room temperature or returned to the incubator.

4. Centrifuge the media samples for 5 min in a microcentrifuge to remove any detached cells. Add an aliquot of the supernatant (5–20%) to aqueous scintillant (2 ml) and count in a scintillation counter for 1–2 min.
5. Repeat using equal numbers of control cells incubated without carbachol or potassium in the culture medium to assess non-regulated secretion.
6. After the last 2 min incubation, the media should be removed and the cells then lysed by a 5-min incubation on ice in either NP40 or NDET (500 µl to 1 ml). The nuclei should be removed from the lysate by centrifugation for 5 min in a microcentrifuge at 14 000 r.p.m. An aliquot of the post-nuclear lysate can then be counted as above to determine the cell-associated counts remaining after stimulation.
7. From the data obtained by following steps **4–6** the total label released in response to stimulation during the 12–18 min can be calculated. A peak of release should occur at the end of the first, second, or third 2-min incubation. Expect on average 20–35% of total cellular label to be released. However, note that regulated secretion from these cells can occasionally be considerably lower or higher than this percentage.

Protocol 5. Regulated and constitutive secretion of marker proteins labelled with [^{35}S]amino acids or [^{35}S]sulphate

Materials

- radiolabelled cells growing at 70–80% confluence that have been plated out on a 10-cm dish at least 2 days previously
- stimulation medium containing 5 mM carbachol or 55 mM KCl (see *Protocol 4*). Note that if the proteins secreted by the cells are to be concentrated by TCA precipitation, the stimulation medium should be serum-free.
- NDET (see *Protocol 4*), supplemented with protease inhibitors where required (1 mM phenylmethylsulphonyl fluoride (PMSF), 10 mM iodoacetamide, 1 µg/ml each of leupeptin, pepstatin, chymostatin, and antipain)
- fixed *Staphylococcus aureus* cells (Cowan strain) as a 50% w/v suspension in PBS
- antibodies to marker proteins

Method

1. Wash labelled cells for 5 min in several changes of culture medium.
2. Replace the culture medium with 5 ml of stimulation medium containing

Protocol 5. *Continued*

5 mM carbachol or 55 mM KCl kept at 37 °C. Leave stimulation medium on for one 10-min interval at 37 °C. Control cells should be treated in parallel with medium lacking the secretagogue in order to assess the level of constitutive release of the protein of interest.

3. Collect the media and place on ice. Spin in microcentrifuge to remove any detached cells or cell debris.
4. Lyse the cells by the addition of 1 ml ice-cold NDET to the dish. Scrape the dish, remove the lysate to a microcentrifuge tube, and leave it on ice for 10 min.
5. Remove nuclei and cell debris by centrifugation for 10 min at 4 °C in a microcentrifuge at 14 000 r.p.m.
6. Concentrate the secretory proteins for analysis. This can be done by trichloroacetic acid (TCA) precipitation in which case the secretory proteins should be collected in serum-free stimulation medium. This is adjusted to 10% TCA and kept on ice for 1 h. The precipitate is then collected by centrifugation for 5 min in a microcentrifuge at 14 000 r.p.m. and dissolved in sample buffer for gel electrophoresis. (Note that at the higher levels of labelling this may not be needed; for a 10-cm dish labelled long-term with 1 mCi and subsequently incubated in 3 ml, 20 μl loaded on to a SDS-gel should give a reasonable signal for the major secreted proteins on a fluorograph after a few days.) If antibodies are available the protein of interest can be immunoprecipitated from the stimulating media (see step **7**).
7. Immunoprecipitate proteins of interest from the cell lysate. It is very difficult to resolve secretory proteins in the cell lysate by electrophoresis, and the proteins usually have to be enriched by immunoprecipitation. Add SDS to the post-nuclear lysate to a final concentration of 0.3%. Pre-clear the lysate by tumbling at 4 °C for 40 min with 50 μl *S. aureus* suspension per 1 ml of lysate, and then remove the matrix using the microcentrifuge (14 000 r.p.m., 5 min). Add the antibody (the amount to be added must be determined empirically) and tumble for 1–18 h. At the end of this period, add enough matrix to bind the antibody, and continue the tumbling for a further 40 min. The complex of matrix/antibody/antigen should then be washed several times with NDET/SDS before analysing by gel electrophoresis. Media samples to be immunoprecipitated should be treated the same way after mixing with an equal volume of 2 × NDET/SDS.
8. Separate proteins by SDS–PAGE and visualize proteins by fluorography.
9. The extent of release in response to stimulation of the protein of interest can then be quantitated. This may be done by counting appropriate excised gel slices in a scintillation counter or by scanning of the

autoradiogram. The amount of the protein in question that is released in response to secretagogue during the 10–15 min stimulation and also the level of constitutive secretion of the same protein during 10–15 min can be calculated as a percentage of total amount of this protein present in the cell. Under normal conditions, 20–35% of a protein stored in the regulated secretory pathway may be released in an experiment of this kind.

5. Cell surface assays for membrane proteins associated with regulated secretion

Cell surface assays may be useful to study membrane proteins thought to be associated with regulated secretory organelle. This is because, after regulated secretory organelles have fused with the cell surface, their membrane proteins will be incorporated into the plasma membrane. Thus, an increase in the cell surface pool of a protein after secretagogue stimulation suggests that the protein is present on the regulated secretory pathway. Also, the plasma membrane may be involved in targeting of some of these membrane proteins. Cell surface assays are described in *Protocols 6 and 7.*

Protocol 6. Cell surface immunoprecipitation

Materials

- an antibody that recognizes the lumenal or extracytoplasmic domain of the membrane protein of interest
- radiolabelled cells growing at 70–80% confluence that have been plated out on a 10-cm dish at least 2 days previously
- stimulation medium containing 5 mM carbachol or 55 mM KCl (see *Protocol 4*)
- NDET supplemented with protease inhibitors where required (1 mM PMSF, 10 mM iodoacetamide, 1 μg/ml each of leupeptin, pepstatin, chymostatin, and antipain)
- fixed *S. aureus* cells (Cowan strain) as a 50% w/v suspension in PBS
- PBS

Method

1. Rinse the cells with culture medium, and replace with stimulation medium supplemented with secretagogue. The stimulation medium should be left covering the cells for 10 min. Control cells should be treated in parallel with medium lacking the secretagogue.

Protocol 6. *Continued*

2. Transfer the dishes on to ice in the cold room, rinse them with ice-cold medium, and then incubate the cells with the antibody at a suitable dilution in ice-cold culture medium for 1–4 h. The antibody diulution must be determined empirically, but should be similar to the dilution used for immunoprecipitation of cell lysates.
3. Wash the dishes on a rotating platform with five changes of culture medium over a period of 1 h.
4. The cells should then be rinsed three times over a period of 5 min with ice-cold PBS.
5. A cell lysate should then be prepared as described in *Protocol 5*. The lysate should be adjusted to 0.3% SDS, and 50 μl *S. aureus* cells added. To help reduce any background, pre-incubate these fixed cells with a post-nuclear supernatant lysate from unlabelled PC12 cells for 1–2 h and then resuspend in NDET/SDS before adding to the radiolabelled lysate. This mix is tumbled for 40 min and then the complex harvested and washed before gel analysis.

Protocol 6 could be adapted for looking at newly synthesized membrane protein appearing on the cell surface during biogenesis in a pulse-chase experiment, or for analysing the effect of secretagogue on the appearance of granule membrane proteins at the plasma membrane in cells that have been labelled for longer periods of time.

Protocol 7. Cell surface iodination

The lactoperoxidase method has been successfully used to iodinate DCG membrane proteins on the surface of PC12 cells. The following is scaled for 10-cm dishes of cells.

Materials

- PC12 cells growing at 70–80% confluence that have been plated out on a 10-cm dish at least 2 days previously
- PBS; PBS/glucose (9 mg/10 ml)
- sodium ^{125}Iodide; lactoperoxidase (25 U/ml in PBS); glucose oxidase (100 U/ml in PBS)
- stimulation medium containing 5 mM carbachol or 55 mM KCl
- NDET supplemented with protease inhibitors where required (1 mM PMSF, 10 mM iodoacetamide, 1 μg-ml each of leupeptin, pepstatin, chymostatin, and antipain)

- fixed *S. aureus* cells (Cowan strain) as a 50% w/v suspension in PBS

Method

1. Rinse the cells for 5 min in several changes of PBS at room temperature.
2. Replace the PBS with 1 ml PBS/glucose that has been supplemented with 0.5 mCi sodium ^{125}Iodide[a], to which should then be added 200 μl of a mix containing 2.5 U lactoperoxidase and 0.5 U glucose oxidase. In the case of dishes to be stimulated, the iodination medium should also contain 5 mM carbachol or 55 mM KCl. The stimulation medium should be left covering the cells for 5 min at room temperature. Control cells should be iodinated in parallel using medium lacking the secretagogue.
3. After 5 min incubation, add a further 100 μl of enzyme mixture and continue the incubation for a second 5-min period. During these incubations gently roll dishes intermittently to ensure that all cells get covered.
4. Gently rinse the cells three times in ice-cold PBS and prepare a lysate for immunoprecipitation as described in *Protocol 5*.
5. Immunoprecipitate with the antibody to a protein of interest.
6. Analyse by gel electrophoresis, and quantitate the results as described in *Protocol 5*.

[a] If cells start to detach from the dish in glucose/PBS (which is common, particularly with cells at high density) gently transfer them to a microcentrifuge tube, and add the sodium ^{125}Iodide and iodination solution with a single addition of 300 μl of the enzyme mixture. The tubes should then be left for 10 min with periodical inversion of the microcentrifuge tubes during the iodination. The cells should then be washed three times by repeated gentle pelleting (about 2000 r.p.m. in a microcentrifuge for 30 sec) followed by resuspension of the cells in PBS. Thereafter preparing the lystate and immunoprecipitation can be performed as described in Protocol 5.

Cell surface immunofluorescence can also be employed to assay membrane proteins on the cell surface. Again the antibody must recognize the lumenal/extracytoplasmic domain of the membrane protein in question. Follow *Protocol 12* using one first-layer antibody but omit the saponin from all stages.

6. Sorting of secretory and membrane proteins to the dense-core granules

Analysis of culture medium and cell lysates as in *Protocols 5–7* can be used to confirm whether a protein is being secreted by either regulated or constitutive secretory pathways. It can also reveal which kind of secretory pathway is being followed by a particular protein. The intracellular sorting of these proteins can also be studied using both biochemical and morphological methods.

6.1 Isolation of DCGs

To demonstrate whether a soluble protein, whose appearance in the medium is regulated, or a membrane protein, whose cell surface appearance is stimulated by secretagogue, are reaching the surface via the DCGs, examination of the organelles characterizing this pathway is necessary. Thus one major requirement for the study of sorting between exocytic pathways is the isolation of DCGs. This can be achieved by sequential sedimentation and density gradient centrifugation using linear Ficoll gradients. Sedimentation gradient centrifugation is the first step in the isolation of DCGs from PC12 cells. In order to be able to monitor the granules after centrifugation, cells should be pre-labelled with either [^{3}H]dopamine or with [^{35}S]methionine (*Protocol 2*) and then a post-nuclear supernatant fractionated on a 1–16% linear Ficoll gradient prepared and used as described in *Protocols 8* and *9*.

Protocol 8. Making linear gradients

Materials

- a gradient maker containing a small magnetic flea in each chamber placed on a magnetic stirrer. The outlet from the gradient maker should lead via a peristaltic pump to the top of a Beckman Ultraclear™ centrifuge tube.[a]
- an overhead stirrer
- a Beckman SW40Ti rotor and an appropriate ultracentrifuge
- buffered sucrose: 0.32 M sucrose/10 mM Hepes pH 7.4
- 40% Ficoll (Ficoll 400; Sigma Chemical Co.) in buffered sucrose (stock solution; this can take 24 h to solvate and it is best to add the solid discontinuously whilst warming the buffer
- 1% Ficoll in buffered sucrose (from the 40% stock)
- 16% Ficoll in buffered sucrose (from the 40% stock)

Method

1. Close the taps to isolate the two chambers on the gradient maker, and add 5.5 ml of 1% Ficoll to the chamber furthest from the point of connection (chamber A) to the peristaltic pump.
2. Add 5.5 ml of 16% Ficoll to the other chamber (chamber B).
3. Switch on the magnetic stirrer and then the pump.
4. Open the tap on the chamber containing 16% Ficoll, then the tap on the chamber containing 1% Ficoll.
5. Collect the resultant gradient in the centrifuge tube observing the following procedures.
   (a) Set the stirrer to mix the solutions evenly.

(b) Set the pump so that a gentle but steady stream of gradient solution enters the centrifuge tube. The aim is to pour as slowly as possible whilst still maintaining a continuous flow.

(c) Let the gradient slide down the wall of the centrifuge tube from the top of the tube.

(d) If an air bubble appears in the section joining the two chambers of the gradient maker, quickly close and then open the tap nearest the bubble and/or tap the gradient maker on the stirrer.

6. Store the gradient upright on a flat surface at 4 °C.
7. Repeat the procedure to make the 16–40% gradient with the following modifications.

   (a) Add 5 ml of each solution to the gradient maker: 16% Ficoll to chamber A and 40% Ficoll to chamber B.

   (b) Stir the 40% solution using both the flea and an overhead stirrer.

[a] Note these procedures have been worked out for a Beckman ultracentrifuge but can be adapted for other systems.

Protocol 9. Sedimentation gradient centrifugation

Materials

- cells growing at 70–80% confluence that have been plated out on a 10-cm dish at least 2 days previously and then labelled with [^{3}H]catecholamine or [^{35}S]methionine
- buffered sucrose; 0.32 M sucrose/10 mM Hepes pH 7
- 1–16% linear gradients
- ball-bearing homogenizer. Use a custom-made ball-bearing homogenizer (EMBL Workshop; see reference 21) or a similar homogenizer from Bernitech Engineering attached to two 2-ml syringes. The ball-bearing should be chosen so as to give a clearance between the barrel and the ball-bearing of 0.012 mm

Method

1. Place the 10-cm dish of cells on ice, and then wash off unincorporated ^{3}H or ^{35}S-label using ice-cold buffered sucrose (three changes within 5 min).
2. The cells should then be scraped with a rubber policeman into 1 ml of the buffered sucrose ready for transfer to a homogenizer.
3. Pre-flush the homogenizer with ice-cold buffered sucrose. The scraped cells should then be transferred to one of the syringes and passed through the homogenizer into the other syringe nine times. The back pressure

Protocol 9. *Continued*

should then be released before removing the homogenate and placing it in a microcentrifuge tube. Investigators may need to optimize the number of homogenization strokes depending on the force they apply. We obtain 98% cell breakage with 95% of the nuclei remaining unbroken using nine strokes and this clearance in the homogenizer.

4. To remove the nuclei and unbroken cells, the homogenate is then centrifuged for 5 min at 11 000 r.p.m. in a microcentrifuge at 4 °C.
5. Gently layer the post-nuclear supernatant on to the top of a 1–16% linear Ficoll gradient[a] (sedimentation gradient) (see *Protocol 8*).
6. Centrifuge at 4 °C for 45 min at 30 000 r.p.m. in a Beckman SW40Ti rotor.
7. Fractionate the gradient collecting 1 ml fractions and store fractions on ice (for short-term storage).
8. To identify DCGs on the sedimentation gradient, count 25–50 μl of each fraction in a scintillation counter (if cells have been pre-incubated with [^{3}H]catecholamine, then a pattern such as that seen in *Figure 3a* should be obtained) or run about 10 μl of each fraction on 10% SDS-gels (if cells have been pre-incubated with [^{35}S]methionine then a pattern such as that seen in *Figure 4a* should be obtained).

[a] Different batches of Ficoll tend to vary. For this reason it is sometimes necessary to change the concentration of the 16% Ficoll solution by plus or minus 1–2%. To determine a suitable percentage, use the [^{3}H]catecholamine distribution as a marker for granules on sedimentation gradients comprised of slightly varying upper percentages of Ficoll. The 1 or 40% Ficoll solutions do not have to be altered.

DCGs, as identified on the sedimentation gradient by the distribution of catecholamine label, can be further purified by equilibrium density gradient centrifugation as described in *Protocol 10.*

Protocol 10. Equilibrium density gradient centrifugation

Materials

- fractions from a sedimentation gradient fractionation (see *Protocol 9*)
- 16–40% linear equilibrium gradients
- Beckman SW40 Rotor and appropriate centrifuge

Method

1. Pool the sedimentation gradient fractions that contain granules as identified by the distribution of catecholamine label after centrifugation (*Protocol 9*).

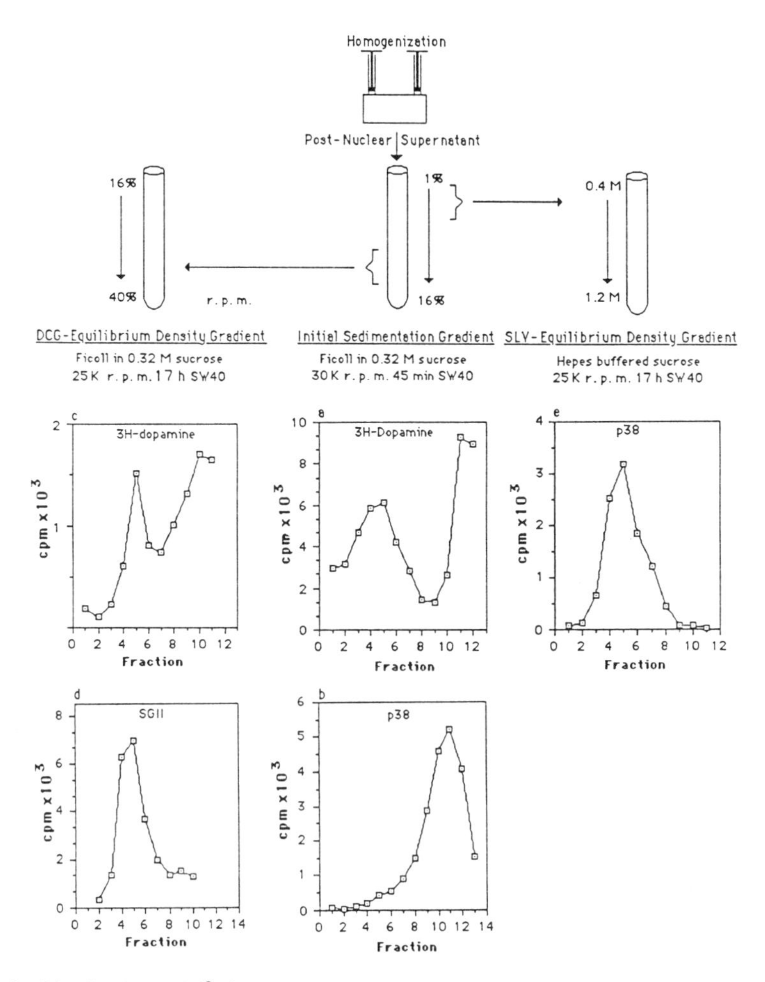

Figure 3. Distribution of [^{3}H]dopamine and radiolabelled proteins on sedimentation, DCG-equilibrium density, and SLV-equilibrium density gradients. 10-cm dishes of PC12 cells were incubated with [^{3}H]dopamine for 1–2 h or with [^{35}S]methionine for 24 h. Cells were rinsed, removed, and homogenized. Post-nuclear supernatants were prepared and then centrifuged on sedimentation gradients. (a) Gradients were fractionated and 50 μl samples counted in a scintillation counter for the distribution of [^{3}H]dopamine or (b) immunoprecipitated with antiserum to a SLV membrane protein (anti-p38 antiserum) and then quantitated. (c) Granule-enriched fractions 3–6 in (a) were then pooled and re-centrifuged on a DCG-equilibrium density gradient, fractionated, and 100 μl samples of each fraction counted in a scintillation counter for the distribution of [^{3}H]dopamine (granules enriched in fractions 4–6) or (d) immunoprecipitated with antiserum against a DCG secretory protein (anti-secretogranin II antiserum) and then quantitated (granules enriched in fractions 4–6). Secretogranin II is SGII. (e) SLV-enriched fractions 10–12 were also pooled from the sedimentation gradient in (b) and were then re-centrifuged on a SLV-equilibrium density gradient, fractionated, and immunoprecipitated with anti-p38 antiserum. They were then quantitated (SLVs enriched in fractions 4–6). The gradient fractions are numbered from the bottom of the gradient.

Protocol 10. *Continued*

2. Layer the pooled fractions on to a 16–40% linear Ficoll gradient (DCG-equilibrium density gradient) (see *Protocol 8*).
3. Centrifuge at 4 °C for 17 h at 25 000 r.p.m. in a Beckman SW40Ti rotor and then fractionate the gradient into 1-ml aliquots.[a]
4. If the cells have been pre-incubated with [^{3}H]catecholamine then, to identify granules on the density gradient (see *Figure 3c*), count 50–100 μl of each DCG-equilibrium density gradient fraction in a scintillation counter. Note, however, that, after 17 h centrifugation, granules may leak catecholamine into the top part of the gradient. If cells have been radiolabelled with [^{35}S]methionine then 20-μl aliquots of each fraction[b] can be analysed by SDS–PAGE to show the presence of granules by the distribution of secretogranin II (see Figure 4b). An alternative marker for the presence of DCGs is ATP (see *Protocol 11*).

[a] After the two-step centrifugation procedure, granules are enriched at least 100-fold and are over 96% free of contaminating organelles such as mitochondria and lysosomes (4).

[b] To determine the distribution of other proteins on the gradients these fractions can be immunoprecipitated or Western blotted (as can fractions from the sedimentation gradients and the SLV-equilibrium density gradients described elsewhere). To immunoprecipitate from gradient fractions, the aliquot to be analysed is diluted 10× in NDET/SDS to 5 ml and then the immunoprecipitation is carried out as described (*Protocol 5*) except that the pre-clearing step is omitted. Because TCA precipitation of samples containing Ficoll and sucrose is difficult, whether or not it will be possible to locate proteins on these gradients by Western blotting will be determined by the abundance of the protein in the cells and the volume of the sample that can be loaded on to the gels being used.

Other methods of isolating the PC12 DCGs have been described and are to be found in references 22–4.

Protocol 11. ATP assay of DCG-equilibrium density gradient fractions

Materials

- crude firefly extract (e.g. Sigma FLE-50)
- 40 mM $MgCl_2$ in water
- 1 M NaOH
- gradient fractions to be analysed
- luminometer (e.g. LKB 1250 luminometer)

Method

1. Add the crude firefly extract to the $MgCl_2$ solution to a final concentration of 5 mg/ml.

2. Thoroughly vortex the extract and keep on ice. Not all the extract will solvate.
3. Add 5 μl of 1 M NaOH to assist solvation, and vortex again.
4. Add 32 μl of solvated extract to 100 μl of gradient fraction.
5. Briefly vortex and immediately monitor luminescence in a luminometer at room temperature.
6. Take readings every second with the gain set high and note the peak reading.[a]
7. For a blank, measure luminescence with 32–64 μl of extract in 100–200 μl buffered sucrose or 16% Ficoll.

[a] The amounts of material used can be increased if the signal is too low for reliable readings to be obtained.

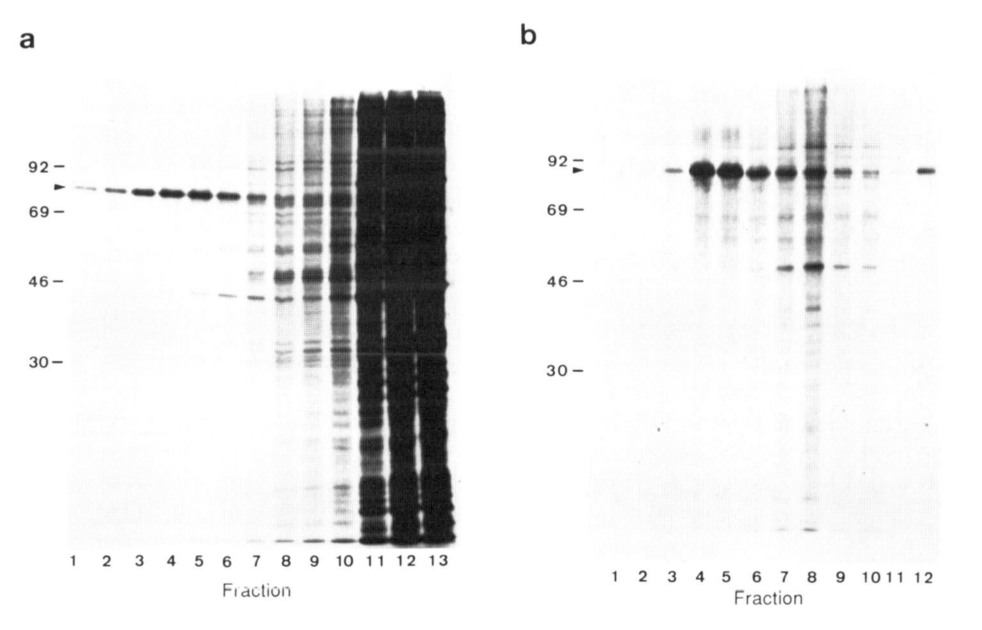

Figure 4. Distribution of radiolabelled proteins on sedimentation and DCG-equilibrium density gradients. 10-cm dishes of cells were labelled with [^{35}S]methionine for 24 h. The cells were rinsed, removed, and homogenized. (a) A post-nuclear supernatant was prepared, then centrifuged on a sedimentation gradient after which 10 μl samples of each fraction were electrophoresed on a 10% SDS-polyacrylamide gel (b). Granule-enriched fractions 3–6 were then pooled and recentrifuged on a DCG-equilibrium density gradient and 20 μl samples of each fraction electrophoresed on a 10% SDS-polyacrylamide gel. The gradient fractions are numbered from the bottom of the gradient. The molecular weights in kilodaltons of marker proteins are indicated. The major granule-associated protein secretogranin II is indicated with an arrowhead.

The profile of ATP usually seen on a DCG-equilibrium gradient is somewhere between the profiles for immunoprecipitated DCG secretory protein and for [^{3}H]dopamine shown in *Figure 3d*.

7. A morphological approach to the localization of proteins on the secretory pathways of PC12 cells

A morphological approach can be used to determine the distribution of proteins being studied with regard to the secretory pathway. In these experiments the aim is to establish the co-localization of the protein of interest with known markers for secretory organelles. One of the more convenient ways of performing this analysis is to use double label immunofluorescence (*Protocol 12*).

Protocol 12. Double label immunofluorescence

Materials

- cells grown on coverslips as described in *Protocol 3*
- 3% paraformaldehyde in PBS pH 7.5, stored as frozen aliquots
- 15 mM glycine/PBS pH 7.5
- wash buffer: PBS/0.5% BSA/0.2% saponin
- mounting medium: 90% glycerol/10% 10 × PBS/3% *N*-propylgallate/0.1% sodium azide. The *N*-propylgallate helps to reduce quenching of the fluorescent signal.
- antibodies to the protein to be localized, and to marker proteins with which to localize the secretory organelles. For this procedure, these must be raised in different species. Thus these antibodies should include, for example, a rabbit polyclonal to the protein of interest and a mouse monoclonal that recognizes a DCG marker such as one of the secretogranin/chromogranins (these latter antibodies are now commercially available). These are the 'primary antibodies' referred to in this protocol.
- second-layer antibodies, which recognize the different primary antibodies and that are conjugated to different fluorophores. An example of a pair of second-layer antibodies that might be used with the primary antibodies suggested above include goat anti-rabbit conjugated to fluorescein and sheep anti-mouse conjugated to rhodamine.

Method

1. Place the coverslips cell-side up in the wells of six-well tissue culture trays. This facilitates subsequent steps. The cells are then gently rinsed in PBS (2–3 changes in 5 min). This and all subsequent steps are carried out at room temperature.

2. The cells are then fixed by incubating the coverslips in paraformaldehyde for 20 min.
3. The paraformaldehyde is then removed by rinsing in PBS, which is followed by quenching twice for 10 min in 15 mM glycine.
4. Rinse in PBS to remove the glycine and then permeabilize the cells by incubation with two changes of wash buffer over a period of 20 min.
5. Prepare the primary antibody mix to be used by diluting them to the desired concentration (to be determined empirically) in wash buffer adjusted to 10% fetal calf serum (FCS), and centrifuging for 10 min in a microcentrifuge at 14 000 r.p.m.
6. Spot out 20–50 μl droplets of the primary antibody solutions on to parafilm in a humidified chamber (to prevent drying-out of the small volume), and, using fine forceps, place the coverslips cell-side down over the antibody solutions. These should then be left for 1 h.
7. The coverslips should then be floated up off the Parafilm by pipetting 200 μl of wash buffer between them and the film. They should then be returned to the six-well dish using forceps and rinsed thoroughly with wash buffer to remove unbound antibody. Six washes of 10 min each is usually sufficient.
8. The cells are then ready for incubation with the second-layer antibodies which should be prepared and used as in steps **5–7**.
9. Finally, the coverslips should be rinsed in PBS (two changes in 10 min), dunked in distilled water, and any remaining fluid drained off with a piece of torn filter paper.
10. The coverslips should then be mounted (cell-side down) on a drop of mounting medium on microscope slides cleaned with 70% ethanol.
11. Nail varnish or warmed dental wax around the coverslip perimeter should be used to fix the coverslip in place.
12. When the nail varnish has dried or the dental wax solidified, gently remove excess fluid from the upper surface of the coverslips with torn filter paper.
13. The cells can then be viewed on a microscope fitted with epifluorescence optics. Using appropriate filters, the same cell can be observed as stained by the two different fluorophores, revealing the location within that cell of the test and marker protein. A pattern match between the two antibodies suggests co-localization of the two proteins within the cell.[a]

[a] A whole series of controls using the first and second antibodies alone and in combination to ascertain both the appropriate concentrations of all the reagents being used, and if any cross-reactivity between the 'wrong' first and second antibodies is occurring, are needed when using this kind of analysis.

8. Sorting of proteins between two regulated secretory organelles

As discused in Section 1, endocrine and neuronal cells have two organelles (SLVs and synaptic vesicles, respectively), which are both thought to be involved in regulated secretion. To aid the study of sorting between SLVs/ synaptic vesicles and DCGs in PC12 cells, SLVs from PC12 cells can be isolated and assayed for the presence of the protein of interest (*Protocol 13*).

Protocol 13. Isolation of PC12 SLVs

Materials

- PC12 cells prepared for fractionation on a Ficoll sedimentation gradient (*Protocol 9*)
- Ficoll sedimentation gradients
- a series of sucrose solutions (made in 10 mM Hepes, pH 7.4) including 1.2, 1.0, 0.8, 0.6, and 0.4 M
- Beckman Ultraclear™ centrifuge tubes, a Beckman SW40 rotor, and suitable ultracentrifuge

Method

1. Prepare and load a 1–16% sedimentation gradient as described in *Protocol 9*.
2. While the gradient is being spun, make a sucrose step gradient for isolating the SLVs. Add 1.8 ml of 1.2 M sucrose (buffered in 10 mM Hepes, pH 7.4) to the bottom of a Beckman Ultraclear™ centrifuge tube with a Pasteur pipette and then consecutively gently layer 1.8 ml of 1.0, 0.8, 0.6, and 0.4 M buffered sucrose into this tube.
3. Pool the top three 1 ml fractions below the lightest (top) fraction from the sedimentation gradient (this is where SLVs are found—see *Figure 3b*) and dilute them with 0.5 ml of Hepes, pH 7.4.
4. Layer the mix on to the top of a sucrose step gradient and centrifuge at 4 °C for 17 h at 25 000 r.p.m.
5. Collect the gradient in 1 ml fractions.
6. The analysis of fractions for the distribution of SLV marker proteins and for other proteins can be performed exactly as for the Ficoll sedimentation and equilibrium gradients described above. A convenient membrane protein marker for the SLVs is synaptophysin or p38 (see reference 4) to which an antibody is commercially available and which can easily be detected by Western blotting of step gradient fractions. It should be found about half-way down the gradient (see *Figures 3e and 5*).

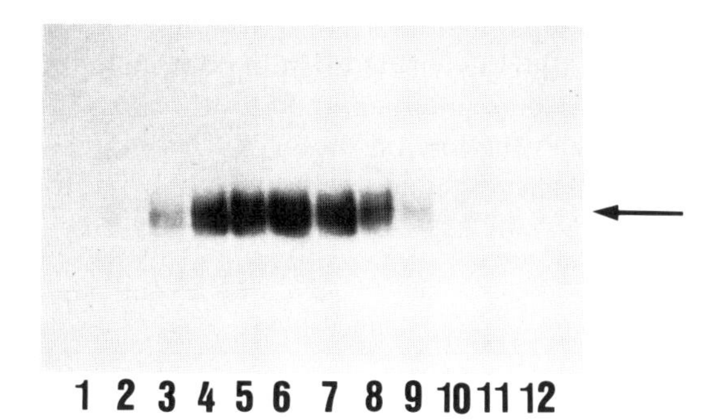

Figure 5. Isolation of SLVs from PC12 cells. A post-nuclear supernatant from a 10-cm dish of homogenized PC12 cells was fractionated on a sedimentation gradient and fractions from the top of this sedimentation gradient containing the bulk of SLVs (identified by the distribution of a SLV membrane protein (p38), i.e. fractions 10–12 in *Figure 3b*) were pooled and loaded on to a SLV-equilibrium density gradient. After centrifugation, samples of each SLV-equilibrium gradient fraction were Western blotted with anti-p38 antiserum and probed with [^{125}I]protein A. p38, marking SLVs, is indicated with an arrow.

9. Generation of new PC12 lines by expression of foreign cDNAs

There are many reasons for wanting to express foreign cDNAs in PC12 cell lines to be used for analysis of membrane traffic, for example, to determine whether the machinery of a rat neuroendocrine cell line can effectively sort foreign proteins between regulated and constitutive secretory pathways (16–18, 25). A second kind of experiment involves the expression of mutants of proteins whose trafficking has been defined. In these cases the experimental aim would be try to identify sequences or domains of the expressed protein that are involved in its targeting. Another use arises where antibodies against the protein of interest were raised against a non-rat protein and do not cross-react well with the PC12 protein but do react well with the product of a cDNA that is available. In this case it may be easier to make a cell line expressing the cDNA than to try to raise antibodies against the rat protein.

Many types of expression vector have proved successful in PC12 cells. These include systems based on recombinant retrovirus infection (25), as well as the use of vectors derived from the simple pSV2 type (16–18). Originally PC12 cells were thought of as hard to transfect but this view has now been modified by the large number of successful attempts at generating new cell lines.

9.1 Transfection

The introduction of DNA into PC12 cells has been performed by a variety of methods. One of the simplest methods is by lipofection. However, most of our cell lines have been obtained using calcium phosphate precipitation of the DNA (*Protocol 14*). We have obtained good levels of expression of a variety of proteins using co-transfection of the SV40 promoter-driven dominant selectable marker for G418 resistance (neomycin phosphotransferase) pSV2-Neo (26) with the cDNA being expressed from a variant vector in which transcription is driven by the Rous sarcoma virus LTR (pRSV, reference 18).

Protocol 14. Transfecting PC12 cells

Materials

- cells that are ready to be passaged
- 10-cm dishes that have been poly-L-lysine coated
- 2 × Hebs and 1 × Hebs: 137 mM NaCl; 5 mM KCl; 5.5 mM glucose; 0.8 mM disodium hydrogen orthophosphate; 21 mM Hepes. The final pH of Hebs should be *exactly* 7.05
- 2 M $CaCl_2$
- G418 ('geneticin')
- the expression vector, and the vector containing the neomycin phosphotransferase (Neo) (where this is not on the expression vector)

Method

1. Plate the cells on to the poly-L-lysine-coated dishes at 3×10^6 cells per 10 cm dish and culture for 2 days.
2. Add 750 μl 2 × Hebs to a sterile 15-ml conical tube.
3. To a second tube, add 100 μg of the expression vector, 20 μg of the neo-containing vector (if separate), 94 μl 2 M $CaCl_2$, and water to 750 μl.
4. Take a 1-ml pipette in an autopipetter and bubble air through the first tube. Do this fast but not furious! Add the DNA mix dropwise to the Hebs to get a fine precipitate that remains initially in suspension.
5. Leave the precipitate at room temperature for 40 min.
6. Rinse the cells twice with 1 × Hebs.
7. Drip the precipitate on to the cells and then leave them at room temperature for 20 min, rocking the dishes occasionally to ensure coverage of the whole surface.
8. Add 10 ml of culture medium and leave the dishes for 5–7 h.
9. Remove the culture medium, and add 1 ml fresh culture medium containing 25% glycerol for exactly 1 min at room temperature.
10. Add 10 ml PBS to the dishes, and then immediately rinse once more

with PBS. Add culture medium and then return the cells to the incubator.

11. The cells are then left for 2 days, split 1:2 on to dishes without poly-L-lysine, and then maintained in 0.5 mg G418/ml culture medium (selecting medium).
12. The cells should then be fed every 2 days for a week, during which almost all of the cells should die and be washed away.
13. Drop the G418 to 0.25 mg/ml (selecting medium) and continue feeding in selecting medium. This is to facilitate the growing up of nascent clones.

9.2 Cloning out cell lines

The transfection outlined in *Protocol 14* generally gives 10–50 clones per dish. These can be seen as small tight clusters of cells that grow to cover several mm by weeks 3 to 6. As soon as these are easily visible they should be picked and regrown in isolation (*Protocol 15*).

Protocol 15. Using cloning rings to isolate transfected lines of PC12 cells

Materials

- cloning rings. These can be cut from stainless steel tubing: i.d. 5 mm, length 8 mm. The ends should be polished after cutting
- glass Petri dish
- Beckman silicone vacuum grease
- dishes of cells in which clones are beginning to appear to the naked eye
- culture medium: trypsin/EDTA solution as used for passaging cells (*Protocol 1*)

Method

1. Stand cloning rings on a thick bed of silicone grease in a glass Petri dish.
2. Autoclave the dish to sterilize the cloning rings.
3. Mark the position of clones to be isolated on the underside of the tissue culture dish.
4. Carefully remove the culture medium from the dish.
5. Using sterile forceps, place the cloning ring around the clump of cells. Press firmly into place to allow the grease to make a watertight seal.
6. Repeat for all the clones to be isolated.

Protocol 15. *Continued*

7. Fill the rings with trypsin solution, and return the dish to the incubator for about 5 min to allow the cells to lift off the dish.

8. Using a Pasteur pipette, remove the cells from the ring and place into the well of a 48-well plate that has been filled with culture medium supplemented with G418 at 0.25 mg/ml.

9. Break up the clump by gentle trituration.

10. Repeat for all the clones and then leave the cells to grow out.

Between 10 and 50 per cent of the clones should express the cDNA when using pRSV in this way. The levels of expression will vary quite widely.

To assay for expression a variety of methods can be used. If a secretory protein has been introduced then medium from the wells in which the clones are growing can be collected and assayed by Western blotting or ELISA. This has the advantage of not being disruptive to the cells and of not requiring the establishment of duplicate series of clones before testing can begin. If a membrane protein has been introduced then duplication of clones will be needed. Assaying for expression by immunofluorescence will require fewest cells and therefore may be performed earlier than other possibilities such as immunoprecipitation or Western blotting of cell lysates.

Once cells expressing the protein of interest have been obtained, the behaviour of those cells as compared to that of the parental line should be checked. Occasionally the selective pressure used to obtain the stably transfected line can produce cells with unusual morphology or with other undersirable characteristics. For the purposes of membrane traffic the presence and functioning of secretory pathways may be the primary parameters to investigate.

References

1. Burgess, T. L. and Kelly, R. B. (1987). *Ann. Rev. Cell Biol.* **3**, 243.
2. Kelly, R. B. (1985). *Science* **230**, 25.
3. De Camilli, P. and Jahn, R. (1990). *Ann. Rev. Physiol.* **52**, 625.
4. Cutler, D. F. and Cramer, L. P. (1990). *J. Cell Biol.* **110**, 721.
5. Greene, L. A. and Tischler, A. S. (1976). *Proc. nat. Acad. Sci., USA* **73**, 2424.
6. Greene, L. A. and Tischler, A. S. (1982). *Adv. Cell. Neurobiol.* **3**, 7373.
7. Winkler, H. (1976). *Neuroscience* **1**, 65.
8. Winkler, H., Apps, D. K., and Fischer-Colbrie, R. (1986). *Neuroscience* **18**, 261.
9. Greene, L. A. and Rein, G. (1977). *Brain Res.* **138**, 521.
10. Greene, L. A. and Rein, G. (1977). *Nature* **268**, 349.
11. Luckenbill-Edds, L., van Horn, C., and Green, L. A. (1979). *J. Neurocytol.* **8**, 493.

12. Tischler, A. S. and Greene, L. A. (1978). *Lab. Invest.* **39**, 77.
13. Wagner, J. A. and Halegoua, S. (1986). *J. Cell Biol.* **103**, 887.
14. Bornstein, M. D. (1958). *Lab. Invest.* **4**, 134.
15. Rosa, P., Hille, A., Lee, R. W. H., Zanini, A., De Camilli, P., and Huttner, W. B. (1985). *J. Cell Biol.* **101**, 1999.
16. Schweitzer, E. S. and Kelly, R. B. (1985). *J. Cell Biol.* **101**, 667.
17. Sevarino, K. A., Felix, R., Banks, C. M., Low, M. J., Montminy, M. R., Mandel, G., and Goodman, R. H. (1987). *J. Biol. Chem.* **262**, 4987.
18. Matsuuchi, L., Buckley, K. M., Lowe, A. W., and Kelly, R. B. (1988). *J. Cell. Biol.* **106**, 239.
19. Tooze, S. A. and Huttner, W. B. (1990). *Cell* **60**, 837.
20. Greene, L. A. and Rein, G. (1977). *Brain Res.* **129**, 247.
21. Balch, W. E., Dunphy, W. G., Braell, W. A., and Rothman, J. E. (1984). *Cell* **39**, 405.
22. Roda, L. G., Nolan, J. A., Kim, S. U., and Hogue-Angeletti, R. A. (1980). *Exp. Cell Res.* **128**, 103.
23. Schubert, D. and Klier, F. G. (1977). *Proc. nat. Acad. Sci., USA* **74**, 5184.
24. Wagner, J. A. (1985). *J. Neurochem.* **45**, 1244.
25. Noel, G., Zollinger, L. Laliberte, F. Rassart, E., Crine, P., and Bioleau, G. (1989). *J. Neurochem.* **52**, 1050.
26. Southern, P. J. and Berg, P. (1982). *J. Mol. appl. Genet.* **1**, 327.

4

Intracellular trafficking in fission yeast

JOHN ARMSTRONG, ERICA FAWELL, and ALISON PIDOUX

1. Introduction

1.1 Genetic approaches to membrane traffic

The analysis of protein targeting and membrane traffic in eukaryotic cells has benefited greatly from the use of genetic methods in the budding yeast *Saccharomyces cerevisiae*. The approach was introduced by Novick and Schekman, whose *sec* mutants define a collection of genes which function at different stages of the secretory pathway. These gene products are now the object of intense study (reviewed in reference 1); already it is clear that the general characteristics of membrane traffic are conserved between yeast and higher eukaryotes, and that several of the *sec* genes have homologues in mammalian cells. Genetic screens in yeast have since been used to identify mutants in other processes of membrane traffic, such as protein targeting to the vacuole (2, 3) and retention of proteins in the endoplasmic reticulum (4). Given the accumulation of biochemical and genetic tools now available, it is likely that *S. cerevisiae* will continue to be popular for studies of different aspects of protein targeting. Several excellent collections of protocols for genetics, cell biology, and biochemistry in budding yeast are now available (e.g. 5, 6).

1.2 Fission yeast versus budding yeast

Despite its numerous attractions, *S. cerevisiae* has limitations as a model for protein targeting in higher eukaryotes. In particular, its membrane systems differ in several significant respects from those of a mammalian cell. Its endoplasmic reticulum resembles that of a higher cell in function but not in morphology. It possesses intermediate secretory compartments analogous to the Golgi complex, but the stacked cisternae characteristic of this organelle are very rarely seen in normal budding yeast. In addition, the biochemistry of the Golgi is very different: glycoprotein modifications such as the addition of galactose and sialic acid do not occur, but instead branched polymers of mannose are added. The vacuole, the equivalent of the lysosome, is present in only one or a few copies per cell, in contrast to the more numerous lysosomes in mammalian cells. In addition, protein targeting to the vacuole does not involve mannose 6-phosphate. Finally, all membrane growth seems to occur in the bud region rather than the mother cell membrane; consequently, all vesicular traffic leaving the Golgi is exclusively targeted to the bud.

The fission yeast *Schizosaccharomyces pombe* is rapidly becoming popular as an alternative genetically tractable eukaryote which is not only phylogenetically distant from *S. cerevisiae* but, in several aspects of its cell and molecular biology, seems more closely to resemble a higher eukaryotic cell. In this respect much interest has recently attached to control of the cell cycle of *S. pombe*, but many other areas are now under study. Of particular relevance to protein targeting is the fact that this organism possesses readily visible Golgi stacks, very reminiscent of those of a higher cell (7), and can add galactose to glycoproteins (8). Since the organism propagates by fission rather than budding, vesicular traffic from the Golgi is clearly not targeted in the same way as in *S. cerevisiae*, although membrane growth is localized to some extent (9).

Genetic analysis of *S. pombe*, while not as advanced as that of *S. cerevisiae*, has progressed to a considerable extent. Likewise, methods for introducing and manipulating genes are developing rapidly. However, most aspects of protein targeting remain to be explored. Likely RNA and protein components of the signal recognition particle have been identified (10–12). Two constitutively secreted proteins (8, 13) and the transferase which adds galactose to them (14) have been characterized. Homologues of the *S. cerevisiae* Sec14p protein (15) and the endoplasmic reticulum protein BiP (known as Kar2p in *S. cerevisiae*) (Alison Pidoux, unpublished) have been found. Several members of the *ypt* family of putative vesicle-targeting proteins have been cloned (16–20). Two of these gene products are closer in sequence to their mammalian than their budding yeast homologues, an observation by no means unprecedented in the molecular biology of this organism.

As with *S. cerevisiae*, the small haploid genome of *S. pombe* makes it amenable to genetic analysis. Its three chromosomes contain approximately 14 million base-pairs of DNA. Several hundred of its genes have been identified (21). Auxotrophic markers are available, of which the most popular are uracil, leucine, and adenine dependence. Haploid cells exist as one of two mating types (discussed more fully in Section 2.4.4). One important difference from *S. cerevisiae* is that diploid cells are generally unstable and will sporulate spontaneously to form haploids. For this reason the basic approaches to complementation analysis and other genetic techniques differ for the two yeasts.

Much of the known biology of *S. pombe* is reviewed in a recent volume

(22). A collection of methods for working with *S. pombe* is also available (23). Here we present some selected protocols for studying the genetics, biochemistry, and cell biology of protein targeting in fission yeast.

2. Isolation of mutants of *Schizosaccharomyces pombe*

2.1 Growth of *S. pombe*

S. pombe is straightforward to grow in rich or defined minimal medium, either in liquid culture or on agar (*Table 1*). Moderate aeration is required. For most purposes strains are grown at 30 °C, at which temperature typical generation times are 2.5 h in rich medium and 3 h in minimal medium depending on the strain. Growth is conveniently monitored by absorbance at 595 nm: a culture of 2×10^7 cells/ml has an absorbance of approximately 1.0 depending on strain type and on the spectrophotometer (this number should be checked with a haemocytometer). For most purposes cells should be used in log phase; saturation occurs at approximately 10^8 cells/ml in rich medium. Patched cultures on agar plates can usually be stored for several weeks at 4 °C. For long-term storage, 1 ml of a fresh liquid culture should be pelleted, resuspended in 1 ml of FM (*Table 1*), frozen rapidly in dry ice/ethanol, and stored at −70 °C. From a frozen culture, cells should be scraped with a sterile spatula or loop, plated for single colonies, and the genotype checked by plating on media which lack supplements as appropriate. The presence of spontaneously formed diploids can be monitored by inclusion of phloxin B(20 mg/l) in plates; haploids are light pink while diploids appear much darker pink in colour due to a higher percentage of dead cells.

2.2 Isolation of temperature-sensitive mutants

As with other micro-organisms, the simplest and most popular phenotype for isolating mutants of *S. pombe* is temperature sensitivity. *Protocol 1* is simple and efficient and can of course be adapted for other selections, such as auxotrophy.

Protocol 1. Isolation of temperature-sensitive mutants of *S. pombe*

Caution! *N*-methyl-*N*′-nitro-nitrosoguanidine is extremely hazardous. It is a potential carcinogen and mutagen, and may explode on prolonged storage. Wear gloves and carry out all manipulations in a fume hood. Before starting experiments, you should consult your local safety adviser and plan a detailed schedule for working with and disposing of this material!

1. From an early stationary phase culture, pellet approximately 10^8 cells at

Table 1. Media for growth of *S. pombe*[a,b]

A. *Media*

YPD		FM	PM (1 litre)[c]		NSM	PMG
Yeast extract	10 g/l	YPD containing 15% (v/v) glycerol	20 × PPN	50 ml	PM with sulphate salts in 50 × salts and 10 000 × minerals replaced by chloride salts	PM with NH_4Cl in 20 × PPN replaced by 1 g/l sodium glutamate
Peptone	20 g/l		50 × Salts	20 ml		
Glucose	20 g/l		Glucose	20 g		
			Vitamins	1 ml		
			Minerals	0.1 ml		
			(20 g Agar + 1 ml 1 M NaOH for solid media)			

B. *Constituents of PM*

20 × PPN	(g/l)	50 × Salts	(g/l)	1000 × Vitamins	(g/100 ml)	10 000 × Minerals	(g/100 ml)
KH phthallate (phthallic acid)	60	$MgCl_2 \cdot 6H_2O$	53.5	Pantothenic acid	0.1	Boric acid	0.5
Na_2HPO_4 (anhydrous)	36	$CaCl \cdot 2H_2O$	0.75	Nicotinic acid	1	$MnSO_4$	0.4
NH_4Cl[d]	100	KCl	50	Inositol	1	$ZnSO_4$	0.4
		Na_2SO_4	2	Biotin	0.001	$FeCl_2 \cdot 6H_2O$	0.2
						Molybdic acid (H_2MoO_4)	0.16
						KI	0.1
						$CuSO_4 \cdot 5H_2O$	0.04
						Citric acid	1

[a] Adenine, uracil, leucine, and other supplements can be made as 1.5 g/l stocks. Use at 75 mg/l (i.e. 50 ml/l) in final media.

[b] Phloxin is used at 20 mg/l.

[c] In making up PM, it is easiest to make up concentrated autoclaved stocks of components and mix when required. Filter-sterilize vitamins and minerals.

[d] For low-nitrogen plate omit NH_4Cl.

Protocol 1. *Continued*

2500 *g* for 5 min in a bench centrifuge. Wash once with 5 ml 0.05 M sodium acetate, pH 5.5.

2. Resuspend in 1.75 ml 0.05 M sodium acetate, pH 5.5, and add 0.25 ml of *N*-methyl-*N*′-nitro-nitrosoguanidine (2 mg/ml in the same buffer). Incubate for 30 min at room temperature with occasional vortexing.
3. Pellet the cells by centrifugation and wash twice with 1 ml PM (see *Table 1*). Allow the cells to recover by incubating at 24 °C for 12 h in 1 ml PM.
4. Plate aliquots of approximately 500 cells/plate on YPD (see *Table 1*). Also plate an aliquot of unmutagenized cells so that the survival rate can be calculated. Incubate at 24 °C until colonies form.
5. Replica-plate and grow at 37 °C and 24 °C. Approximately 50% of mutagenized cells should survive, of which approximately 1% should fail to grow or grow poorly at 37 °C. Survival rates are most easily altered by changing the length of time of exposure to the mutagen.

2.3 Enrichment for density mutants by Percoll gradient centrifugation

Novick and co-workers (see reference 1) observed that *S. cerevisiae* mutants defective in the secretory pathway become denser than wild-type at the restrictive temperature, since biosynthesis continues but membrane growth is halted. This was used as an effective enrichment for secretory mutants. The authors have adapted this method to isolate secretion mutants of *S. pombe* (A. Tsun and J. Armstrong, unpublished). Dense cells are enriched by centrifugation through Percoll, a form of colloidal silica coated with polyvinylpyrrolidone. Since increased density is likely to be a characteristic of cells defective in numerous different aspects of protein targeting, the protocol is given here.

Protocol 2. Percoll gradient centrifugation

1. To pre-form Percoll gradients, make 85% Percoll (Pharmacia) in 0.1 M NaCl, and centrifuge 12 ml portions for 30 min, 16000 r.p.m. in a Sorvall SS-34 rotor or equivalent. The resulting gradients should range from 1.00 to 1.14 g/ml; this can be checked by incorporating marker beads (Pharmacia).
2. Mutagenize 3×10^7 cells as described in *Protocol 1*, allow to recover for 12 h at 24 °C in PM.
3. Pellet the cells, resuspend in fresh PM, and incubate at 37 °C for 3 h.
4. Pellet the cells again, resuspend in 0.5 ml 0.1 M NaCl, and load gently on

Protocol 2. *Continued*

top of the pre-formed gradient. Centrifuge for 5 min at 5000 r.p.m. in a swing-out rotor in a standard bench centrifuge.

5. Remove the top 10.5 ml from the gradient, which includes the cells of normal density (1.095 g/ml). Plate aliquots of the remainder of the gradient on YPD-agar (see *Table 1*), and incubate at 24 °C until colonies appear.
6. Replica-plate and incubate at 24 °C and 37 °C to identify temperature-sensitive cells.

2.4 Back-crossing and recombination analysis

Mutants of interest should be back-crossed, preferably at least three times, to ensure that the phenotype is the consequence of a single mutation. To do this, mix a loopful of mutant cells and one of an unmutagenized strain of opposite mating type in a drop of sterile water on a low-nitrogen agar plate (see *Table 1*) to form a 1 cm^2 patch. This medium encourages mating and sporulation of the resulting zygotes. Isolate spores by glusulase treatment as described in *Protocol 12*, plate and incubate at 24 °C until colonies form, and then identify temperature-sensitive progeny by replica plating. If the phenotype is due to a single mutation half the progeny will be temperature-sensitive. A smaller fraction of temperature-sensitive progeny indicates that the phenotype is a consequence of multiple mutations. If the mutagenesis was performed with cells of only one mating type, progeny from back-crossing are a convenient source of mutants of the opposite mating type.

Since diploids are generally unstable, the assignment of mutants to complementation groups is most easily accomplished by recombination analysis. Cross each mutant with the others and analyse the progeny by tetrad analysis as described in *Protocol 13* (random spore analysis as described in *Protocol 12* can also be used). If the mutations are unlinked, one of each tetrad will be wild-type. If all progeny display the mutant phenotype the two mutants belong in the same complementation group.

3. Biochemical and morphological approaches to intracellular trafficking

The principal attraction of working with yeasts rather than higher eukaryotes is their amenability to genetic analysis. For questions of protein targeting, however, the next stage in general will be to study the phenotypic consequences of mutations and the behaviour of gene products using the approaches of cell biology. Such methods have tended to lag behind their equivalents or higher cells, partly as a consequence of the small size and tough cell wall of yeasts. Nevertheless, these approaches are now being developed in *S. pombe*. Some of those already available will now be described.

3.1 Markers for different compartments

For fluorescence microscopy, a variety of probes can be used to identify specific cellular compartments (some by vital staining) in fission yeast (see *Table 2*). Other markers used in budding yeast (5) are also likely to work. Traffic through the secretory pathway can be monitored using invertase or acid phosphatase (8, 13). Antibodies have been described for acid phosphatase and a simple solution assay in available for the enzyme activity (13). A list of solutions is described in *Table 3*.

Table 2. Compartment markers

Cellular component	**Compound**
Cell wall	Calcofluor
Mitochondria	Rhodamine 123
F-actin	Rhodamine-phalloidin
Nuclear and mitochondrial DNA	DAPI (4,6′-diamidino-2-phenylindole)

Table 3. Solutions

TE	10 mM Tris-HCl, pH 8; 1 mM EDTA
SP1	1.2 M sorbitol; 50 mM sodium citrate ($Na_3C_6H_5O_7{\cdot}H_2O$); 50 mM sodium phosphate ($Na_2HPO_4{\cdot}7H_2O$); 40 mM EDTA (ethylenediamine tetraacetate). pH to 5.6; autoclave
PEM	100 mM Pipes, pH 7; 1 mM EGTA (ethyleneglycol-bis-(2-aminoethyl ether) *N,N,N′,N′*-tetraacetic acid); 1 mM $MgSO_4$
PEMS	PEM with 1.2 M sorbitol
PEMF	PEM with 0.5% (w/v) fish skin gelatine (Sigma)
Poly-L-lysine	1 mg/ml in distilled water
RIPA	0.5% (v/v) Nonidet P40; 0.5% (v/v) sodium deoxycholate; 50 mM Tris-HCl, pH 7.5; 150 mM NaCl; 1 mM EDTA. *Optional*: 1 g/l SDS
TNE	50 mM Tris-HCl, pH 7.5; 150 mM NaCl; 1 mM EDTA. *Optional*: 1% (v/v) Triton-X114
SDS sample buffer	20 g/l SDS (sodium dodecyl sulphate); 50 mM Tris-HCl, pH 6.8; 2 mM EDTA; 10% (v/v) glycerol; 0.3 mg/ml bromophenol blue; 2% (v/v) 2-mercaptoethanol
HE	50 mM Hepes (*N*-2-hydroxyethylpiperazine-*N′*-2-ethanesulphonic acid), pH 7.9; 5 mM EDTA; 100 mM NaCl
HK	200 mM Hepes, pH 7.9; 10 mM EDTA; 200 mM NaCl; 20 g/l SDS; 200 μg/ml proteinase K added just before use
Phenol/chloroform	Melt phenol at 50 °C. Extract three times with an equal volume of 1 M Tris-HCl, pH 8.0. Add: 1 volume chloroform; 1/25 volume isoamyl alcohol; 1 g/l 8-hydroxyquinoline
10 mM dNTPs	10 mM each of dATP, dCTP, dGTP, and dTTP
5 × RT	500 mM Tris-HCl, pH 8.3; 700 mM KCl; 50 mM $MgCl_2$; 50 mM dithiothreitol
10 × PCR buffer	500 mM KCl; 100 mM Tris-HCl, pH 8.2; 15 mM $MgCl_2$
Buffer I	20 mM Tris-HCl, pH 7.5; 2 mM EDTA; 200 mM LiCl
Buffer II	40% PEG 4000; 100 mM LiCl in TE. Sterilize by filtration

3.2 Immunofluorescence microscopy

The principal barrier to immunofluorescence of intracellular proteins in yeast is the cell wall, which must be removed to allow access of antibodies. For uncharacterized combinations of antigen and antibody, different fixation conditions should be tested. It is also not uncommon for an antiserum to react to yeast components (e.g. carbohydrate) unrelated to the specific antigen; therefore antibodies should be affinity-purified or other evidence sought for their specificity.

The following procedures are modified from Hagan and Hyams (24). All incubations should be carried out at room temperature with gentle agitation.

Protocol 3. Immunofluorescence of *S. pombe* cells

1. Grow cells to mid-log phase (1×10^7 cells/ml) and fix using either aldehydes or organic solvents. Unless stated otherwise, the following manipulations are carried out at room temperature, with cells re-suspended at 5×10^7 cells/ml.
   (a) *Paraformaldehyde fixation.* To prepare paraformaldehyde. Dissolve paraformaldehyde (4% w/v) in PEM (see *Table 3*) by heating to 80 °C, with stirring, in a fume hood. Cool, filter, and store in aliquots at −20 °C. Thawed aliquots should not be refrozen. Spin down 1 ml cells. Wash cells with PEM and fix in 4% (w/v) paraformaldehyde in PEM for 60 min at room temperature.
   (b) *Formaldehyde/glutaraldehyde fixation.* Add 1/10 volume of freshly prepared 30% paraformaldehyde in PEM followed after 30 s by 1/250 volume of 50% glutaraldehyde and agitate for 30–90 min.
   (c) *Methanol fixation.* Filter an *S. pombe* culture on to Whatman GF/C glass-fibre filters (up to 30 ml per 2.4-cm filter). Shake collected cells into methanol at −20 °C and incubate for 8 min.
2. After fixation, spin down the cells, wash them twice with PEM, once with PEMS (*Table 3*), and resuspend at 5×10^7 cells/ml in PEMS.
3. Digest the cell walls with 0.4 mg/ml Zymolyase-20T (Seikagaku Kogyo Co.) for 30–60 min at 37 °C. Check for formation of spheroplasts by mixing with 10 μl 10% SDS and observing by phase-contrast microscopy; spheroplasted cells appear dark. Cells grown in different media, e.g. YPD (*Table 1*), may require more enzyme, a longer incubation time, or a mixture of enzymes, for example, Zymolyase-20T and Novozym 234 (Novo Enzyme Products).
4. Wash cells three times in PEMS; permeabilize cells with 1% (v/v) Triton X-100 in PEMS for 30 sec.

5. Wash once with PEMS and twice with PEM; incubate cells for 60 min in PEMF (*Table 3*) to block non-specific binding sites.
6. Incubate cells overnight with primary antibody (try 1/100 of a previously uncharacterized serum), wash three times with PEMF, and incubate for 4 h with an appropriate fluorophore-conjugated second antibody.
7. Wash cells three times with PEMF, resuspend at 5×10^7 cells/ml, apply to a poly-L-lysine-coated coverslip and mount in PBS/glycerol/anti-fade (e.g. Citifluor; Agar Aids).

3.3 Electron microscopy

The following protocol has been selected since it results in good definition of membrane structures.

Protocol 4. Permanganate fixation for electron microscopy

1. Grow *S. pombe* cells to a density of approximately 10^7 cells/ml. Centrifuge 1 ml for 2 min at 13 000 *g* in microcentrifuge at room temperature.
2. Wash the cells three times with 1 ml of distilled water.
3. Resuspend in 1 ml of freshly prepared 20 g/l $KMnO_4$ in water for 30 min at room temperature with frequent shaking or mixing.
4. Wash the cells three times in 1 ml of water, resuspending the pellet each time by vortexing. On the final wash pack the cells by brief centrifugation at 13 000 *g*, rotate the tube through 180 °C, and respin. This forms a compact pellet.
5. Dehydrate the pellet through a graded series of alcohols:
   - once in 70% ethanol for 15 min
   - once in 90% ethanol for 15 min
   - twice in 100% ethanol for 20 min
   - once in propylene oxide for 20 min
6. Embed using Epon or equivalent (e.g. Taab 812; Taab Laboratories) according to the manufacturer's instructions and section with a diamond knife. Sections can be post-stained with lead citrate (25) and viewed by electron microscopy.

3.4 *In vivo* labelling and immunoprecipitation

This section describes the labelling of *S. pombe* proteins with [^{35}S]compounds and subsequent processing for immunoprecipitation. Since proteins involved in membrane trafficking may be membrane-associated, a protocol is included for separation of hydrophobic proteins using the detergent Triton X-114 (26). *S. pombe* can be labelled with either [^{35}S]sulphate or [^{35}S]methionine. [^{35}S]sulphate, however, besides being cheaper, offers the advantage of quick incorporation into both cysteine and methionine.

Protocol 5. Radiolabelling of *S. pombe*

1. (a) Grow *S. pombe* overnight in PMG (*Table 1*). Harvest the cells, resuspend at 2×10^7 cells/ml in PMG, and add [^{35}S]methionine (Amersham) at about 100 μCi/ml.

   (b) Alternatively, grow cells to log-phase in NSM (*Table 1*) plus 200 μM ammonium sulphate; then resuspend in NSM (*Table 1*) and add 50–100 μCi/ml [^{35}S]sulphate (Amersham) per 10^7 cells.

   Cells can be labelled for up to two generation times.
2. Harvest the cells by centrifugation, wash once in distilled water, and resuspend in the same volume of SP1 (*Table 3*) containing 0.4 mg/ml Zymolyase 20T. Incubate at 37 °C for 30–60 min.
3. Harvest spheroplasts by brief centrifugation. Then proceed with either SDS lysis or Triton X-114 extraction.

SDS lysis

1. Resuspend in 1% SDS in 50 mM Tris-HCl pH 6.8 at 2×10^7 cells/ml. Vortex the sample, boil for 5 min, and then centrifuge for 5 min.
2. Dilute the supernatant with 9 volumes of RIPA without SDS (see *Table 3*) at 4 °C and then process for immunoprecipitation as described in *Protocol 6*.

Triton X-114 extraction

1. Resuspend spheroplasts in 1 ml cold 1% Triton X-114 in TNE (*Table 3*), mix by vortexing, and incubate on ice for 5 min. Centrifuge at 4 °C for 5 min to remove insoluble material and remove the supernatant to a clean tube.
2. Warm the samples to 30 °C for 3 min and centrifuge for 1 min at room temperature. Remove the upper aqueous phase to a clean tube and add 50 μl of Triton X-114 (to 0.5% (v/v)). Add 900 μl TNE to the lower detergent phase and incubate both fractions at 4 °C for 2–3 min.
3. Warm both phases to 30 °C for 3 min and centrifuge for 1 min at room temperature.
4. Recover the respective re-extracted detergent and aqueous phases. Add TNE or Triton X-114, respectively, to obtain equal volumes and approximately the same salt and detergent content in both samples.

Samples made by SDS lysis or Triton X-114 extraction are immunoprecipitated according to *Protocol 6*.

Protocol 6. Immunoprecipitation

1. To a sample in RIPA or Triton X-114/TNE add an appropriate dilution of antibody and incubate on ice for 60 min.
2. Add 30 μl of a 50% slurry of protein A-Sepharose (Pharmacia) in TNE (previously blocked by incubation for 1 h in 1% (w/v) bovine serum albumin (BSA)). Incubate overnight at 4 °C with gentle mixing.
3. Centrifuge for 30 sec to pellet the beads, discard the supernatant, and wash the beads three times in TNE.
4. Add 30 μl of SDS-sample buffer, boil for 5 min, and analyse by SDS–PAGE.

---

4. Molecular genetics in *S. pombe*

S. pombe genes can be cloned by complementation of genetic defects and also using standard molecular biological methods. Cloned genes can be manipulated *in vitro* and reintroduced into *S. pombe*. Several of the molecular genetic techniques developed for *S. cerevisiae* can now be used in *S. pombe*. These include gene disruption and replacement, cloning of mutant alleles by gap-repair, and expression of genes from plasmids, including inducible expression. However, transformation efficiencies and the ratio of homologous to non-homologous recombination tend to be lower in *S. pombe*. Also, plasmid copy number and stability are less well controlled than in *S. cerevisiae*. Due to the relatively large size of *S. pombe* centromeres (27), single-copy plasmids equivalent to 'CEN' vectors are not available; therefore any genetic approach requiring stable single copies of genes necessitates integration into the genome.

This section briefly reviews strategies which can be used to clone *S. pombe* genes and gives methods for transformation and gene disruptions.

4.1 Cloning and genome analysis

The *S. pombe* genome is approximately 14 megabases, comprising three chromosomes of 5.7, 4.6, and 3.5 megabases (28). Classical genetics has assigned approximately 300 genes to positions on the three chromosomes (21) and contributions to physical mapping are now being made by molecular biology. *Not*I and *Sfi*I restriction maps of the *S. pombe* genome have been made using pulsed-field gel electrophoresis. Genes can be assigned to particular *Not*I or *Sfi*I fragments by Southern blotting (28).

The relatively small genome size means that a reasonably low number of clones in a genomic library need to be screened to find a particular gene (for instance a phage lambda library of 10 kb insert size should contain five genome equivalents in 7000 clones). Cloning by cross-hybridization using a probe from another species can be a useful way of isolating *S. pombe* homologues. The best choice of probe is not necessarily a sequence from *S. cerevisiae* since several *S. pombe* genes cloned to date have a greater degree of homology to the corresponding mammalian genes than to these from *S. cerevisiae*. If available, sequences from both mammalian and *S. cerevisiae* sources should be used. Information on homology may best be exploited by designing degenerate oligonucleotides corresponding to regions of the protein conserved between other species. Codon usage studies for *S. pombe* have been made and these may be useful when designing oligonucleotides (29). Such oligonucleotides can be used in a polymerase chain reaction and for screening libraries.

Analysis of cloned genes has enabled consensus sequences for *S. pombe* introns to be determined (30). The 5′ splice site consensus is GTANG, generally followed by 20–100 bases before the branch site, NNCTPuAN, then 3–16 bases to the 3′ splice site consensus NAG (usually AAG or TAG). In contrast to *S. cerevisiae*, introns are common in nuclear genes of *S. pombe*.

Although not covered in this chapter, cloning by complementation of mutations is a commonly used method for isolating genes and for identifying extragenic suppressors. In addition, *S. pombe* homologues of specific genes have been cloned by complementation of *S. cerevisiae* mutations (see reference 31). This approach requires a *S. pombe* cDNA library in a suitable *S. cerevisiae* vector.

4.2 Complementary DNA (cDNA) isolation by the polymerase chain reaction

A cDNA version of the gene may be required for construction of fusion proteins in *Escherichia coli* when making antibodies, for *in vitro* transcription–translation systems or simply to confirm the assignment of introns. If the gene is to be expressed in other species (for instance for complementation of mutations in *S. cerevisiae*, or for localization studies in mammalian cells) it is preferable to use the cDNA as *S. pombe* intron splice sites may not be recognized in other species.

Isolation of the cDNA by the polymerase chain reaction (PCR) (32) avoids the need to screen a cDNA library and is quicker and easier. The first step in cDNA isolation is to make total RNA from *S. pombe* (*Protocol 7*). (The authors have not found it necessary to isolate polyA$^+$ RNA.) cDNA is then synthesized using reverse transcriptase and oligo-dT (*Protocol 8*), and the cDNA used as the template for PCR amplification using specific oligonucleotide primers (*Protocol 9*).

4.2.1 Choice of oligonucleotide primers

Oligonucleotides, corresponding to the 3′ and 5′ ends of the gene, should have approximately 20 bases of complementarity to the target sequence. Pairs of oligonucleotides which have a significant degree of complementarity to each other should be avoided, as should oligonucleotides with 'unusual' sequences such as long runs of purines or predicted secondary structure. Restriction enzyme sites can be added at the 5′ end of each oligonucleotide to facilitate cloning of the fragment after amplification. Choice of restriction site depends on which vectors the cDNA is to be cloned into and which sites are not present in the open reading frame. The efficiency of cutting PCR products is increased if, in addition to the restriction site, another few bases (about three) are added at the 5′ end of the oligo.

Protocol 7. RNA preparation (adapted from reference 33)

All solutions and plastics should be autoclaved. It is important to wear gloves during all manipulations as ribonucleases are present on the hands.

1. Grow a 20 ml culture to 10^7 cells/ml.
2. Pellet the cells at 2500 *g* for 5 min, resuspend in 2 ml ice-cold water, and transfer to a 2-ml round-bottomed tube. Spin down and drain the pellet thoroughly.
3. Resuspend in 50 µl ice-cold HE buffer (see *Table 3*) and add 0.2 ml of baked acid washed glass beads (425–600 µm; Sigma). Vortex for at least 2 min.
4. Wash the beads three times with 100 µl cold HE, and pool the supernatants in a clean tube.
5. Centrifuge the extract for 30 sec at 13 000 *g*, room temperature. Remove 300 µl supernatant into a tube containing 225 µl HK buffer (including proteinase K, see *Table 3*).
6. Incubate at 37 °C for 1 h.
7. Add 0.5 ml phenol/chloroform (see *Table 3*), mix, and centrifuge for 5 min.
8. Take off the aqueous phase into a fresh tube, add 50 µl 3 M sodium acetate, pH 5 and 1 ml 100% ethanol. Incubate on dry ice for 1 h or overnight at −20 °C.
9. Centrifuge for 10 min at 13 000 *g*. Wash the pellet with 70% ethanol and dry it.
10. Resuspend in 100 µl sterile water. Read the optical density at 260 nm (OD_{260}): 1 unit = 40 µg. This method yields ~ 200 µg RNA.

Protocol 8. cDNA synthesis

1. Denature 10–20 μg RNA in ~ 50 μl water by heating to 85 °C for 5 min; then plunge into ice.
2. Add the RNA to the other ingredients in an Eppendorf tube so that the final volume is 80 μl, and contains:
   - 16 μl 5 × RT (*Table 3*)
   - 80 units placental ribonuclease inhibitor (e.g. RNasin; Promega)
   - 8 μl 10 mM dNTPs (*Table 3*)
   - 8 μg oligo-dT (e.g. 12–18mer; Pharmacia)
   - 20 units AMV reverse transcriptase (e.g. Gibco-BRL)
3. Incubate at 42 °C for 90 min.
4. Precipitate using 1/10th volume of 3 M sodium acetate, pH 5 and 2 volumes ethanol. Centrifuge for 5 min at 13 000 *g*. Resuspend in 20 μl water.

Protocol 9. Polymerase chain reaction

1. Mix:
   - 5 μl cDNA, from *Protocol 8*
   - 1/10 volume of 10 × PCR reaction buffer (*Table 3*)
   - 250 μl 10 mM dNTPs (*Table 3*)
   - 100 pmol 5′oligonucleotide
   - 100 pmol 3′oligonucleotide
   - 2.5 units Taq polymerase (Perkin-Elmer Cetus)
   - distilled H_2O to 100 μl

   Overlay with one drop of light mineral oil (Sigma).
2. Perform the PCR reaction. Choice of conditions depends on the oligonucleotides and on the length of sequence to be amplified. For amplification of a 1 kb fragment using oligonucleotides with 20 bases of complementarity to the target sequence the following conditions generally work:
   (a) Denature at 92 °C for 1 min.
   (b) Anneal at 50 °C for 1 min.
   (c) Extend at 70 °C for 2 min.

   After repeating steps (a)–(c) for 25 cycles:
   (d) Final exension at 70 °C for 5 min.

More cycles will increase the amount of product but will also increase the likelihood of incorrect nucleotides being incorporated; 25 cycles should give sufficient material to clone.

3. When the reaction is complete analyse 10 μl of the reaction on an agarose gel to confirm that a fragment of the expected size has been amplified.
4. Remove the other 90 μl from beneath the mineral oil and transfer to a clean tube, extract with phenol/chloroform, and precipitate with ethanol.
5. Cut the DNA with the relevant restriction enzymes and gel purify. Ligate into a suitable vector.
6. cDNA sequences made by PCR should be sequenced to check that no errors have been made by the Taq polymerase.

4.3 Introduction of plasmids into *S. pombe*

It may be desirable to introduce plasmids into *S. pombe* for several reasons. Expression of cloned genes on multicopy plasmids may facilitate immunolocalization, especially if the endogenous expression level is low. If no antibodies to the protein are available the gene can be engineered to include sequence encoding an epitope for which monoclonal antibodies are available, for example, the 9E10 epitope from the myc oncogene product (34). Foreign genes and specifically mutated versions of endogenous genes can also be expressed from plasmids.

4.3.1 *S. pombe* plasmids

Plasmids used in *S. pombe* contain yeast sequences and sequences which allow selection (e.g. ampicillin resistance) and replication in *E. coli*. Selectable marker genes which are commonly used include the *S. cerevisiae LEU2* gene which complements the *S. pombe* $leu1^-$ mutation, and the *S. cerevisiae URA3* and *S. pombe* $ura4^+$ genes which complement $ura4^-$ mutations. Another alternative is the *ade6-704/sup3-5* system (35).

A plasmid containing only a marker gene and bacterial sequences can replicate in *S. pombe* up to 200 copies per cell (in contrast to *S. cerevisiae* where such plasmids are not maintained without integration); however the transformation efficiency is very low. Transformation efficiency is greatly improved when *S. cerevisiae* 2 μm or *S. pombe ars* sequences (e.g. *ars1*) are included on the plasmid. Copy number is reduced by the inclusion of these elements to 5–10 per cell for 2 μm and 15–80 per cell for *ars* plasmids. Both types of plasmid are mitotically unstable due to asymmetric segregation between daughter cells; under non-selective conditions the plasmid is very rapidly lost from the population (typically less than 5% still contain the plasmid after 20 divisions). Even under selective conditions less than half the cells may actually contain the plasmid. Symmetric segregation appears to be improved by inclusion of the *stb* element in conjunction with *ars1*. This element also improves transmission of plasmids in meiosis which is otherwise very inefficient. The *stb* element does not function in an analogous way to the CEN elements used in *S. cerevisiae* vectors (36). It is also possible to integrate plasmids into the *S. pombe* genome (23).

Plasmids in *S. pombe* can be subject to some degree of rearrangement. Inclusion of the *ars* element appears to reduce this problem. Plasmids often exist in a partially polymeric form which can make their recovery into *E. coli* difficult (37), such as when isolating complementing plasmids from a library.

A selection of available *S. pombe* plasmids is listed in *Table 4*. Genes may be expressed from their own or from heterologous promotors. The *S. pombe* alcohol dehydrogenase (ADH) promotor in the vector pEVP11 gives a very high level of expression (38). If overexpression of a gene is likely to be deleterious to the cell, vectors containing the SV40 early promotor (39) which gives a moderate level of expression can be used. Plasmids containing inducible promotors have recently been developed for *S. pombe*:

- FPB promotor (40). Expression is repressed by 8% glucose and de-repressed in low glucose and 3% glycerol. Regulation occurs over a 100-fold range and intermediate levels of expression can be achieved with different carbon sources. Cells must be maintained in log phase, since de-repression occurs in stationary phase.
- Glucocorticoid system (41). The rat glucocorticoid receptor is introduced into *S. pombe* on a LEU2 plasmid and constitutively expressed from the *S. pombe* ADH promoter. The cDNA of interest is introduced on a URA3 plasmid containing the *S. cerevisiae* CYCl promoter and multiple glucocorticoid response elements. Expression from this promoter is normally low. Addition of steroids such as dexamethasone induces expression 20–70 fold.
- *nmt* promoter (42). The *nmt* promoter (no message in thiamin) has been incorporated into replicating and integration plasmids. Expression of

Table 4. Some *S. pombe* plasmids

Plasmid	Elements	Marker	Promoter	Expression	Reference
pfl20	*ars1 + stb*	*URA3*	–		48
pIRT2	*ars1*	*LEU2*	–		49
WH5	2 μm	*LEU2*	–		50
pEVP11	2 μm	*LEU2*	*adh*	High	38
pSM+/–	2 μm	*LEU2*	SV40 early	Moderate	39
pART1	*ars1*	*LEU2*	*adh*	High	51
pMB332	*ars1 + stb*	*URA3*	*adh*	High	52
pCHY21	*ars1 + stb*	*URA3*	*fbp*	Inducible	40
p2UG	2 μm	*URA3*	*CYCI*	Inducible	41
pREP	*ars1*	*LEU2/ura4*	*nmt*	Inducible	42

heterologous genes is at a very high level in normal minimal media and is repressed almost completely by 2 μM thiamine.

Transformation protocols for *S. pombe* have been adapted from *S. cerevisiae* methods. A quick and easy method which is adequate for most purposes utilizes lithium salts (chloride or acetate) and polyethylene glycol. A method for LiCl transformation (43) is given in *Protocol 10*. If it is important to get as many transformants as possible, for example, when transforming a mutant strain with a gene bank, the spheroplast (44), spheroplast and lipofection (45), electroporation (46), or high-efficiency spheroplast (47) methods can be used.

Protocol 10. LiCl transformation

1. Grow a culture to saturation in YPD or minimal medium; use 10 ml of this culture to inoculate 40 ml of fresh medium. Grow for 4–5 h.
2. Centrifuge at 2500 *g* for 5 min, wash once in sterile water, and resuspend in 0.6 ml Buffer I (*Table 3*), giving a final volume of ~ 1.2 ml and a concentration of ~ 2×10^9 cells/ml.
3. Incubate at 30 °C for 1 h with gentle agitation.
4. Divide into 0.2 ml portions in Eppendorf tubes and add 0.1–1 μg DNA in 10 μl TE (*Table 3*) to each. Incubate at 30 °C for 30 min without shaking.
5. Add 0.7 ml Buffer II (*Table 3*) to each tube. Incubate at 30 °C for 30 min without shaking.
6. Heat shock for 25 min at exactly 46 °C.
7. Spread on to minimal selective plates and allow to dry before inversion. Colonies appear in 4–6 days at 30 °C. Typical transformation efficiency for plasmid DNA is $1\text{–}5 \times 10^3$ per μg.

4.4 Gene disruption

Gene disruption (knock-out) experiments allow the phenotype of a null allele to be analysed and thereby determine whether or not the gene is essential. A construct is made *in vitro* in which the open reading frame (ORF) of the gene is replaced by a selectable marker leaving flanking sequences of the gene on either side. These sequences direct homologous recombination when the linear fragment is transformed into cells. If it is suspected that the gene of interest does not encode an essential function the disruption can be performed in a haploid strain. Otherwise, gene disruptions are done in diploid cells, in which one chromosomal copy of the gene is replaced by the selectable marker and the other remains intact. The disruptants are then sporulated and the progeny analysed for co-segregation of the marker and a phenotype such as lethality.

4.4.1 A note on strains

S. pombe has two mating types, plus (+) and minus (−). Cells of opposite mating type can mate with each other, under conditions of nutrient starvation, forming a diploid zygote which undergoes meiosis and sporulation to give four haploid progeny. Homothallic strains (designated h^{90}) can switch mating type every other generation and can therefore mate within the population. There are also heterothallic (h^+ and h^-) non-switching strains which are self-sterile. In h^{90} strains the mating type locus consists of three main elements. The expression locus *mat1* contains either plus or minus information which is copied from the silent *mat2* or *mat3* cassettes, respectively. Heterothallic strains contain deletions and/or rearrangements at the mating type locus. Commonly used heterothallic strains are h^{+N} and h^{-S} though there are many other alleles. h^{-S} strains contain a deletion of the *mat2* ('plus' information) and are therefore stable. h^{+N} has a rearrangement at the mating type locus such that only 'plus' information is expressed; this state can revert to h^{90} at 1 in 10^4. (For a more comprehensive discussion of this topic see reference 53.)

Because haploidy is the preferred state most h^+/h^- strains are unstable and sporulate spontaneously after a few days' growth on minimal media. To maintain cells as diploids until sporulation is required a strain can be used which contains two different *ade6* mutant alleles, *ade6–210* and *ade6-216* (54). These alleles show intragenic complementation so a diploid strain containing both of them, but neither on its own, is able to grow on media lacking adenine. Because the mutations are in the same gene, reversions to ade6+ through recombination are very infrequent. The procedures described in the following sections use this type of strain.

An alternative is to use an h^{+N}/h^{+N} diploid which is stable because sporulation requires expression of both plus and minus specific genes. The h^{+N} allele reverts to h^{90} at a frequency of 1 in 10^4. Individual cells will then be h^{+N}/h^+ or h^{+N}/h^-. h^{+N}/h^- cells can sporulate; sporulating colonies are easy to identify because they turn brown when exposed to iodine vapour. Alternatively, h^{+N}/h^{+N} diploid disruptants can be mated to h^-/h^- diploids forming a tetraploid zygote which will give rise to some h^{+N}/h^- diploids which can themselves be sporulated.

4.4.2 Construction of the 'knock-out fragment' and transformation into *S. pombe*

A construct is made in which as much as possible of the ORF is replaced by the gene for a selectable marker (see *Figure 1a*). At least 500 bp of upstream or downstream sequence should be left on either side; a kilobase is preferable. A marker commonly used for this purpose is the *S. pombe ura4* gene. Strains to be transformed with such a construct should have the *ura4-*

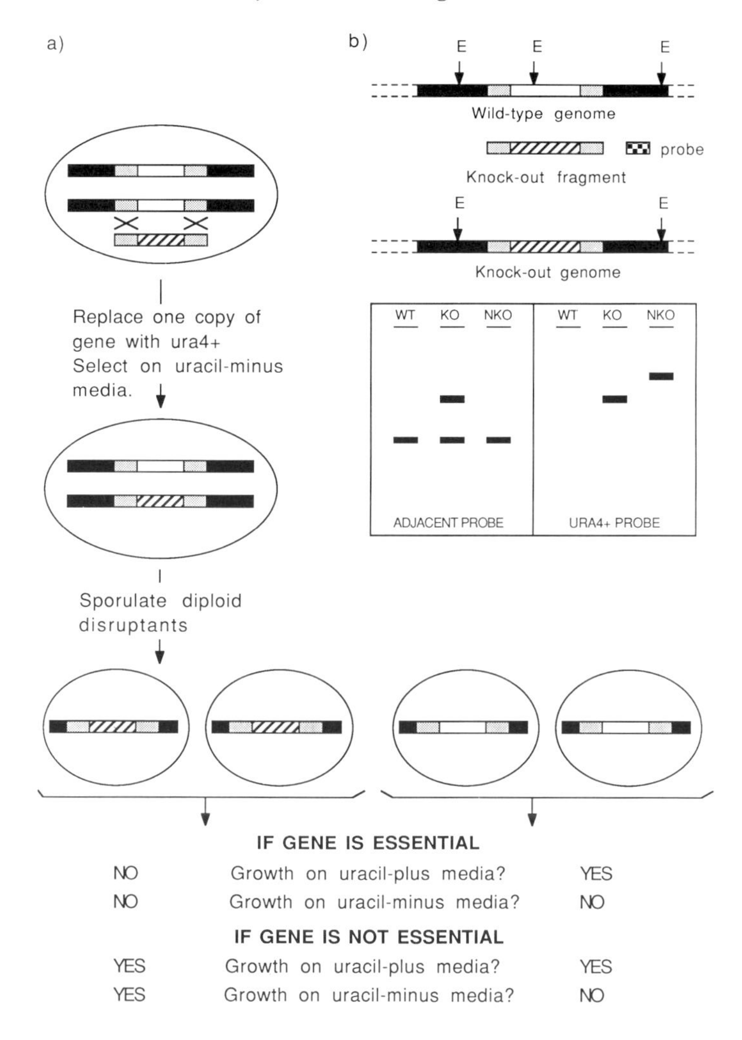

Figure 1. (a) Schematic representation of gene disruption. Gene is represented by open boxes, *ura4* gene by diagonal stripes, flanking sequences by stippled shading, and other genomic sequences by black boxes. Diploid cells are transformed with the knock out fragment which integrates by homologous recombination to replace one copy of the gene. Transformants are selected on media lacking uracil. Disruptants are sporulated and replica plated on to the media indicated to determine whether the disrupted gene is essential. (b) Selection of disruptants by Southern blotting. Genomic DNA from transformants and the parental strain (wild-type) is cut with the a restriction enzyme (E). A probe (chequered box) made from a region outside the knock-out fragments is hybridized to the DNA on a Southern blot. One hand hybridizes in wild-type (WT) and non-disruptant (NKO) DNA. The wild-type band and a novel band hybridize in the disruptant (KO) DNA. When a *ura4* probe is used, no band hybridizes in the wild-type DNA, the novel band hybridizes in the disruptant DNA, and a hand of random length hybridizes in the non-disruptant DNA.

D18 allele which is a deletion of the entire *ura4* gene, as opposed to smaller deletions such as *ura4-D6* or point mutations. This prevents integration at the *ura4* locus. The *S. cerevisiae LEU2* gene can also be used but transformants may grow slowly due to the gene being present in only one copy. The *S. cerevisiae URA3* gene cannot be used for this purpose as it will only complement the *ura4-* mutation when present on a multicopy plasmid.

When the construct has been made the relevant piece should be cut out of the cloning vector and 1 μg purified as a linear fragment for transformation. The LiCl transformation method described above for plasmids is adequate for this purpose, though the transformation efficiency may be low. If transforming an *ade6-210/ade6-216 ura4-D18/ura4-D18* h^+/h^- strain with a fragment containing the *ura4* gene, plate on to media lacking uracil and adenine. Again, colonies should appear in about 4 days.

4.4.3 Selection of disruptants

Transformants must be analysed by Southern blotting for correct integration of the fragment. The frequency of homologous integration varies and disruptants may be between 1 and 100% of transformants. Genomic DNA from the transformants and the parental wild-type strain is digested with restriction enzyme(s) chosen to give a different-sized fragment for the wild-type and disrupted gene (e.g. an enzyme which cuts within the marker gene but not in the original ORF, as shown in *Figure 1b*). This DNA is subjected to gel electrophoresis, blotted, and hybridized to a probe made from outside the region used for the 'knock-out' fragment but internal to the fragment cut by the enzyme(s).

Pick several transformants and streak to single colonies on selective plates. Patch a single colony from each transformant on to a fresh plate and use this for DNA preparations and sporulation.

Protocol 11. *S. pombe* genomic DNA preparation (adapted from reference 55)

All procedures are carried out at room temperature, unless stated otherwise.

1. Grow up a culture in 5 ml of selective media for 2–3 days.
2. Pellet the cells at 2500 *g* for 5 min and resuspend in 0.25 ml SP1 (*Table 3*) containing 0.4 mg/ml Zymolyase-20T. Incubate at 37 °C for 30–60 min, until spheroplasts form. Cell wall digestion can be checked by placing 10 μl of cells on a slide and adding 1 μl 10% SDS: spheroplasted cells look black.
3. Centrifuge the spheroplasts for 30 sec. Resuspend in 0.5 ml of TE, add 50 μl 10% SDS, and mix well. Add 165 μl 5 M potassium acetate and incubate on ice for 30 min.

4. Centrifuge for 10 min at 13 000*g*. Take off the supernatant and add to 0.75 ml cold isopropanol. Incubate on dry ice for 5 min and then centrifuge for 10 min at 13 000 *g*.
5. Dry the pellet and resuspend it in 300 μl TE (*Table 3*). Add 1 μl DNase-free RNase (10 mg/ml) (Gibco-BRL), and incubate for 30 min at 37 °C.
6. Add 300 μl phenol/chloroform. Vortex and centrifuge 5 min at 13 000 *g*.
7. Take off the aqueous phase into a clean tube, add 30 μl 3 M sodium acetate and 750 μl ethanol, incubate for 10 min on dry ice, centrifuge for 10 min at 13 000 *g*, and wash with 70% ethanol.
8. Dry the pellet and resuspend it in 20 μl TE.
9. Digest 5–10 μl with relevant restriction enzymes and analyse by agarose gel electrophoresis, including DNA from the parental strain as a control. Perform a Southern blot and probe (56). Disruptants will contain the wild-type band and an extra band of a size predicted by the genomic restriction map. Non-disruptants will contain only the wild-type band (see *Figure 1b*). Integration of the fragment at the correct site can be confirmed by stripping the filter (56) and reprobing it with the marker gene, e.g. *ura4*. In this case only the novel band should hybridize; the wild-type will lack the band and transformants in which the marker has integrated at another locus will contain a hybridizing band of random length.

Transformants may also be analysed for correct integration of the fragment by PCR amplification of genomic DNA. For example, an oligonucleotide priming in the marker gene together with one outside the region used in the 'knock-out fragment' will give no product for wild-type and irrelevant integrants, while disruptants will give a product of predictable size.

4.4.4 Sporulation and spore analysis

Once a disruptant has been identified it can be sporulated to determine whether the gene disruption is lethal or has some other phenotype. Spores can be analysed using tetrad dissection or by random spore analysis. Tetrad analysis involves the microdissection of the four-spore ascus and requires a microscope fitted with a micromanipulator, specially pulled glass needles, and patience to learn the technique. At least 10–20 tetrads should be dissected. Spores are germinated and grown up separately; replica plating is then used to determine the genotype of the progeny. Alternatively, random spore analysis can be used. This is technically simpler—liberated spores are plated on rich plates and then replica-plated on to various media to determine their genotype—but it lacks the genetic rigour of tetrad dissection which is able to analyse the progeny from single diploid cells.

For both methods, the first step is to patch the disrupted diploid strain on to low nitrogen plates (containing adenine if using *ade6-210/ade6-216 strain*) followed by incubation at 24 °C for 2 or 3 days. The presence of asci can be shown by exposing the colonies to iodine vapour—hold the inverted plate over a few iodine crystals for 1 min (extended exposure will kill cells). Colonies containing asci turn brown, due to storage of amylose in the spores. The presence of asci can also be determined microscopically: asci are rod-shaped and contain four phase-bright spores which look like pearls.

Protocol 12. Random spore analysis

1. Scrape a loopful of sporulating cells (it does not matter if the asci walls are breaking down) into 1 ml of sterile water containing 10 µl glusulase (Du Pont) or 2 µl helicase (IBF Biotechnics). These are crude snail enzymes that break down yeast cell walls, causing vegetative cells to burst. The enzymes also break down asci walls but leave spores intact. Incubate overnight at 25–29 °C.
2. Wash twice in sterile water. Count the spores with a haemocytometer and plate out at 500 per plate on YPD plates. Incubate at 24 °C for a few days until colonies form.
3. Replica plate on to various media. For the example strain, required media would be:
   - PM + ade + ura: all colonies will grow
   - PM + ade: only ura$^+$ colonies grow. This will be half the colonies if the disrupted gene is not essential, or none if the disrupted gene is essential.
   - PM (*Table 1*): this is to check that no diploids survived the glusulase/ helicase treatment. The diploids will grow on PM; the haploid progeny will not. YPD + phloxin (*Table 1*) can also be used for this purpose—diploid colonies are dark pink on phloxin plates while haploids are pale pink.

Protocol 13. Tetrad analysis

1. Take a loopful of cells from a patch in which the asci are not breaking down (e.g. after 1–2 days on sporulating media), mix with sterile water on a YPD plate, and streak cells as thinly as possible in a line on one side of the plate.
2. Place the plate on the stage of the dissecting microscope. Dissect out asci and place them in a line 1 cm away from the streak at 5 mm apart from each other.

3. Incubate at room temperature (20–22 °C) overnight, or for a few hours at 30–35 °C (unless there are *ts* alleles involved). This will cause asci walls to break down (glusulase/helicase is not used).
4. Dissect out the spores placing each set of four 5 mm apart in a line perpendicular to the streak. If the spores have germinated and divided, do not use that ascus since it is hard to be certain of the origin of each cell.
5. Incubate at 24 °C for 3 days until colonies form. If the disrupted gene is essential two of four spores from each ascus will give rise to colonies, and these will both be ura^-, as can be shown by replica plating on PM + ade. If the disrupted gene is not essential all four spores will form colonies of which two will be ura^+ and two will be ura^-.

   Similarly, if other markers are involved, replica-plate on to appropriate media.

4.4.5 Gene replacements

Genes which have been altered *in vitro* can be reintroduced into the genome in an analogous way to gene disruption. A disrupted diploid strain (or a disrupted haploid strain, if the gene is not essential) in which one chromosomal copy of the gene has been replaced by *ura4* is transformed with a linear fragment containing the altered gene and flanking sequences. Cells are plated on media containing 5-fluoro-orotic acid (5-FOA, an analogue of a precursor of uracil) and uracil (57). Cells which have an intact uracil biosynthesis pathway die. Colonies form from ura^- cells in which integration has occurred at the correct locus. If a diploid strain was used it is then sporulated. Half the progeny will contain the altered gene.

References

1. Schekman, R. (1985). *Ann. Rev. cell. Biol.* **1**, 115.
2. Bankaitis, V. A., Johnson, L. M., and Emr, S. D. (1986). *Proc. nat. Acad. Sci., USA* **83**, 9075.
3. Rothman, J. H. and Stevens, T. H. (1986). *Cell* **47**, 1041.
4. Semenza, J. C., Hardwick, K. G., Dean, N., and Pelham, H. R. B. (1990). *Cell* **61**, 1349.
5. Guthrie, C. and Fink, G. R. (ed.) (1990). *Methods in enzymology*, Vol. 194. Academic Press, London.
6. Tartakoff, A. M. (ed.) (1989). *Methods in cell biology*, Vol. 31. Academic Press, London.
7. Smith, D. G. and Svoboda, A. (1972). *Microbios* **5**, 177.
8. Moreno, S., Ruiz, T., Sanchez, Y., Villanueva, J. R., and Rodriguez, L. (1985). *Arch. Microbiol.* **142**, 370.

9. Johnson, B. F., Miyata, M., and Miyata, H. (1989). In *Molecular biology of the fission yeast* (ed. A. Nasim, P. Young, and B. F. Johnson), pp. 331–66. Academic Press, London.
10. Ribes, V., Dehoux, P., and Tollervey, D. (1988). *EMBO J.* **7**, 231.
11. Liao, X., Brennwald, P., and Wise, J. A. (1989). *Proc. nat. Acad. Sci., USA* **86**, 4137.
12. Poritz, M. A., Siegel, V., Hansen, W., and Walter, P. (1988). *Proc. nat. Acad. Sci., USA* **85**, 4315.
13. Schweingruber, M. E., Fluri, R., Maundrell, K., Schweingruber, A.-M., and Dumermuth, E. (1986). *J. Biol. Chem.* **261**, 15877.
14. Chappell, T. G. and Warren, G. (1989). *J. Cell Biol.* **109**, 2693.
15. Salama, S. R., Cleves, A. E., Malehorn, D. E., Whitters, E. A., and Bankaitis, V. A. (1990). *J. Bacteriol.* **172**, 4510.
16. Fawell, E., Hook, S., and Armstrong, J. (1989). *Nucl. Acids Res.* **17**, 4373.
17. Fawell, E., Hook, S., Sweet, D., and Armstrong, J. (1990). *Nucl. Acids Res.* **18**, 4264.
18. Miyake, S. and Yamamoto, M. (1990). *EMBO J.* **9**, 1417.
19. Hengst, L., Lehmeier, T., and Gallwitz, D. (1990). *EMBO J.* **9**, 1949.
20. Haubruck, H., Engelke, U., Mertins, P., and Gallwitz, D. (1990). *EMBO J.* **9**, 1957.
21. Kohli, J. (1987). *Curr. Genet.* **11**, 575.
22. Nasim, A., Young, P., and Johnson, B. F. (ed.) (1989). *Molecular biology of the fission yeast.* Academic Press, London.
23. Moreno, S., Klar, A., and Nurse, P. (1990). In *Methods in enzymology*, Vol. 194 (ed. C. Guthrie and G. R. Fink), pp. 795–823. Academic Press, London.
24. Hagan, I. M. and Hyams, J. S. (1988). *J. Cell Sci.* **89**, 343.
25. Reynolds, E. C. (1963). *J. Cell Biol.* **17**, 208.
26. Bordier, C. (1981). *J. Biol. Chem.* **256**, 1604.
27. Clarke, L., Amstutz, H., Fishel, B., and Carbon, J. (1986). *Proc. nat. Acad. Sci., USA* **83**, 8253.
28. Fan, J.-B., Chikashige, Y., Smith, C. L., Niwa, O., Yanagida, M., and Cantor, C. R. (1988). *Nucl. Acids Res.* **17**, 2801.
29. Sharp, P. M., Cowe, E., Higgins, D. G., Shields, D. C., Wolfe, K. H., and Wright, F. (1988). *Nucl. Acids Res.* **16**, 8207.
30. Mertins, P. and Gallwitz, D. (1987). *EMBO J.* **6**, 1757.
31. Fikes, J. D., Becker, D. M., Winston, F., and Guarente, L. (1990). *Nature* **346**, 291.
32. Saiki, R. K., Gelfand, D. H., Stoffel, S., Scharf, S. J., Higuchi, R., Horn, G. T., Mullis, K. B., and Erlich, H. A. (1988). *Science* **239**, 487.
33. Kaufer, N. F., Simanis, V., and Nurse, P. (1985). *Nature* **318**, 78.
34. Munro, S. and Pelham, H. R. B. (1987). *Cell* **48**, 899.
35. Hottinger, H., Pearson, D., Yamao, F., Gamulin, V., Cooley, L., Cooper, T., and Soll, D. (1982). *Mol. Gen. Genet.* **188**, 219.
36. Heyer, W. D., Sipiczki, M., and Kohli, J. (1986). *Mol. cell. biol.* **6**, 80.
37. Hagan, I., Hayles, J., and Nurse, P. (1988). *J. Cell Sci.* **91**, 587.
38. Russell, P. and Nurse, P. (1986). *Cell* **45**, 145.
39. Jones, R. H., Moreno, S., Nurse, P., and Jones, N. C. (1988). *Cell* **53**, 659.
40. Hoffman, C. S. and Winston, F. (1989). *Gene* **84**, 473.

41. Picard, D., Schena, M., and Yamamoto, K. R. (1990). *Gene* **86**, 257.
42. Maundrell, K. (1990). *J. Biol. Chem.* **19**, 10857.
43. Broker, M. (1987). *Biotechniques* **5**, 516.
44. Beach, D. and Nurse, P. (1981). *Nature* **290**, 140.
45. Allshire, R. C. (1990). *Proc. nat. Acad. Sci., USA* **87**, 4043.
46. Hood, M. T. and Stachow, C. (1990). *Nucl. Acids Res.* **18**, 688.
47. Okazaki, K., Okazaki, N., Kume, K., Jinno, S., Tanaka, K., and Okayama, H. (1990). *Nucl. Acids Res.* **18**, 6485.
48. Losson, R. and Lacroute, F. (1983). *Cell* **32**, 371.
49. Hindley, J., Phear, G., Stein, M., and Beach, D. (1987). *Mol. cell. Biol.* **7**, 504.
50. Wright, A. P. H., Maundrell, K., Heyer, W. D., Beach, D., and Nurse, P. (1986). *Plasmid* **15**, 156.
51. McLeod, M., Stein, M., and Beach, D. (1987). *EMBO J* **6**, 729.
52. Broker, M. and Bauml, D. (1989). *FEBS Lett.* **248**, 105.
53. Egel, R. (1989). In *Molecular biology of the fission yeast* (ed. A. Nasim, P. Young, and B. F. Johnson), pp. 31–73. Academic Press, London.
54. Beach, D., Rodgers, L., and Gould, J. (1985). *Curr. Genet.* **10**, 297.
55. Durkacz, B., Beach, D., Hayles, J., and Nurse, P. (1985). *Mol. Gen. Genet.* **201**, 543.
56. Sambrook, J., Fritsch, E., and Maniatis, T. (1989). *Molecular cloning* (2nd edn). Cold Spring Harbor Laboratory Press, Cold Spring Harbor, New York.
57. Grimm, C., Kohli, J., Murray, J., and Maundrell, K. (1988). *Mol. Gen. Genet.* **215**, 81.

5

Network antibodies as tools for studying intracellular protein traffic

DAVID VAUX

1. Introduction

Many questions in cell biology can be reduced to a study of the interactions between two protein components, whether these are enzyme and substrate, surface receptor and extracellular ligand, or targeting signal and intracellular receptor. Many biochemical techniques can be applied to these problems, but there are occasions when these techniques are insufficient. Under these circumstances, it is justified to turn to an immunological approach to modelling the interaction between two proteins and to make use of the ability of the immune system to generate connected networks of interrelated immunoglobulins, *network antibodies*. In this chapter I will briefly outline the theoretical basis for this approach, consider the advantages and limitations of the method, and then provide detailed practical protocols.

1.1 Network theory

An immunoglobulin (Ig) molecule consists of a basic heterotetrameric unit containing two identical heavy chains and two identical light chains linked by disulphide bonds. Each chain is encoded by a separate gene and is organized into domains: a chain consists of one variable region and one or more constant domains. The antigen-combining site is assembled from the variable domains of both heavy and light chain, with the predominant contacts occurring between antigen and three hypervariable regions within the variable domain, known as the complementarity determining regions (CDRs). A central concept in immunology is that the antigen-combining site of an Ig is a novel structure which may itself be immunogenic (1). The epitopes of an Ig variable region are described as *idiotopes*, the array of idiotopes on an immunoglobulin constitutes its *idiotype*, and antibodies elicited against them are described as *anti-idiotope* or *anti-idiotype* antibodies. The humoral response to a syngeneic immunoglobulin contains only anti-idiotype antibodies which fall (in the simplest analysis) into two classes—those recognizing idiotopes which lie wholly within the antigen-combining site of the first immunoglobulin, and those recognizing idiotopes which lie wholly outside this region. The latter give rise to 'framework' anti-idiotypes, while the former give rise to 'internal image' anti-idiotypes, so called because the antigen-combining site of the anti-idiotype antibody is related spatially to the original antigen. This network of interconnected antibodies has many implications for our understanding of diversity and the control of the immune system (1–4).

1.2 Structural mimicry by anti-idiotype antibodies

Anti-idiotype antibodies are immunoglobulins, similar in structural organization to any other antibody. While there are many solved crystal structures for immunoglobulins, including anti-idiotype Igs, there are few solved structures of Ig–antigen complexes. Our understanding of the way in which an immunoglobulin 'fits' to the epitope which it recognizes is far from complete.

There is a single example of a solved structure of an idiotype–anti-idiotype complex for which the idiotype–antigen complex is also solved (5). We do not know whether the interaction between an anti-idiotype antibody and its antigen, the idiotype Ig, differs in any *general* way from the interaction of an antibody with a non-Ig antigen. This is a very important point, since we are forced to assume that the same constraints which govern the size and accessibility of non-Ig epitopes are extendable to idiotopes. This is an explicit assumption underlying the use of internal image anti-idiotype antibodies for probing protein–protein interactions.

By definition, idiotype Ig is recognized by an internal image anti-idiotype only when the anti-idiotype Ig antigen combining site mimics the starting antigen used to elicit the idiotype antibody. Thus, if the starting antigen is a 'ligand', one can seek to identify the 'receptor' by virtue of its affinity for the anti-idiotype antibody, and this is the basis for the experimental approach described here.

1.3 Anti-idiotype antibodies as probes of protein–protein interactions

Sege and Peterson (6) were the first to show that internal image anti-idiotype antibodies could be exploited for the analysis of protein–protein interactions. If an antibody recognizes a specific ligand, then a subset of its internal image anti-idiotypes will mimic the ligand and bind specifically to its receptor. Thus it is possible to produce antibodies which define previously unidentified receptors starting from a knowledge of the ligand alone, using the terms ligand and receptor in their most general sense. A wide ranging survey of this technology is found in *Methods in enzymology*, Volume 178 (7). Internal image anti-idiotype antibodies which usefully mimic ligand will be referred to simply as anti-idiotypes in the rest of this chapter.

1.4 Limitations of the anti-idiotype antibody approach

There are a number of requirements which may limit the utility of the anti-idiotype approach. These fall into three groups: first, it is essential that the ligand of the interaction under study constitutes an epitope recognized by the immune system used; second, it is necessary that the idiotype produced elicits an internal image anti-idiotype antibody in the second-round immunization; and, third, the mimicry of the original antigen by the internal image anti-idiotype must not be limited by distant structural features of the immunoglobulin molecule itself.

The first of these requirements is very important, and not amenable to prior testing. It is not sufficient that any part of the ligand be recognized; the exact site of the interaction must form an immunogenic epitope. Only antibodies to this epitope will be relevant idiotypes for generation of anti-idiotype reagents which recognize the 'receptor' for the ligand. It follows that only signals which are small enough to be mimicked by an antibody binding site will be amenable to analysis using the network antibody approach. Unfortunately, it is not always possible to be sure that a given idiotype is relevant even if a functional assay is available, because an irrelevant antibody that recognizes a nearby epitope may interfere with the functional assay by a steric hindrance. The strongest suggestion that the idiotype is relevant occurs if the Fab fragment inhibits the functional assay.

If the murine immune system is indeed 'complete' in terms of network theory then any idiotype immunoglobulin will generate anti-idiotypes (8), so the second requirement is probably met during the second-round immunization.

Even if the first two requirements are met and true internal image anti-idiotype antibodies result, there may be other constraints which prevent these from being useful mimics of the original antigen. The potential for an epitope to give rise to a useful internal image anti-idiotype obviously depends upon the ability of the CDRs to reproduce the shape of this original epitope. One may imagine that an epitope ranges in complexity from a simple structure which may be copied using a variety of primary amino acid sequences, to a structure so unique that the only way in which it may be recreated is by the repetition of exactly the same amino acid sequence. Thus, one would predict that the CDRs of true internal image anti-idiotype antibodies should range from perfect sequence homology with the underlying epitope in some cases, to no apparent sequence homology in others, although the three-dimensional structure of the epitope is still being copied by the anti-idiotype in each case. Presumably those anti-idiotypes requiring perfect sequence convergence to generate the internal image of the epitope will arise with a lower frequency than those in which mimicry may be achieved with a variety of primary sequences.

The other structural assumption which is made when protein–protein interactions are modelled with anti-idiotype antibodies is that the constraint imposed upon the anti-idiotype CDRs by the Ig framework will not prevent the whole molecule from being a useful mimic of the starting antigen. For example, a difficulty occurs if the original ligand is recognized by a binding site buried inaccessibly within a narrow cleft; in this circumstance the entire methodology may work perfectly to produce an internal image anti-idiotype antibody which exactly mimics the ligand, but the resulting reagent will be unable to recognize the 'receptor' if the cleft is too narrow to admit an immunoglobulin molecule.

For all of these reasons, the use of an anti-idiotype approach to study protein–protein interactions should not take precedence over conventional biochemical approaches, but should be regarded as an adjunct to these methods.

2. Strategy for anti-idiotype production

The strategy employed for anti-idiotype production depends on the information available about the system under study. The fundamental strategic decision depends on whether or not an assay exists to permit the selection of the relevant idiotype antibody for use as antigen. If such as assay exists, then it is straightforward in principle to prepare an inhibitory monoclonal antibody. This antibody is then purified and used as the antigen for generating anti-idiotype reagents. Under these circumstances, the identification of anti-idiotypes is reasonably simple: putative anti-idiotypes should interact specifically with the idiotype (and this interaction should be inhibited by the original antigen) and should also independently inhibit the assay. This approach is described in Section 2.1.

If no assay is available for selection of a single relevant idiotype, then a mixture of first round idiotype antibodies is used as the antigen for the generation of anti-idiotypes. In this situation the selection of anti-idiotypes is not simple: the straightforward specificity checks involving idiotype competition are not available and no assay is available for functional inhibition experiments. Putative anti-idiotypes are initially identified by indirect evidence based on information available about the interaction under study, such as the expected intracellular location, tissue distribution, or cell cycle regulation of the antigen. Candidate anti-idiotypes may then be tested by the generation of third-round antibodies (anti-anti-idiotype antibodies; Ab 3), which should recreate the specificity of the original idiotype and recognize the original antigen. This approach is described in Section 2.2.

2.1 Sequential immunization using a defined idiotype

This sequential approach is experimentally straightforward and success depends on the exact ligand being sufficiently immunogenic to produce an appropriate idiotype, which is itself immunogenic enough to result in a usable internal image anti-idiotype. It requires a functional assay for the interaction under study which can be used to select appropriate monoclonal idiotype antibodies for use as antigen in the second round immunization.

Although this approach was pioneered using outbred rabbits for immunization there is a potentially serious problem of allotype responses complicating the analysis of the anti-idiotype antiserum. It is probably better to make use of inbred animal strains to avoid allotype differences and mice have been successfully used. We have found that the best anti-idiotype response is usually obtained by immunization of the syngeneic mouse strain with fixed idiotype-secreting hybridoma cells. It is also possible to generate anti-idiotype responses by growing hybridomas in non-irradiated syngeneic animals, although this is not optimal because the animals rarely survive long enough for isotype switching to occur and the titres and affinities of the IgM produced are not usually high.

Table 1 lists the reagents and equipment required for the protocols described in this chapter, together with information about suppliers where the source is important. *Protocol 1* covers fixation and immunization with idiotype-secreting hybridoma cells for generation of a polyclonal anti-idiotype response. If a response is observed then the mice may be used to obtain monoclonal anti-idiotype antibodies.

Protocol 1. Immunization with fixed idiotype-secreting hybridoma cells

1. Grow hybridoma cells secreting the relevant idiotype in their usual growth medium. Harvest the cells by centrifugation (400 *g* for 5 min at room temperature). Wash the pelleted cells twice with phosphate-buffered saline (PBS) and count using a haemocytometer or automatic counter. Aim to immunize with 5×10^6 to 1×10^7 cells per mouse.
2. Resuspend the cells in 1 ml of PBS and add 1 ml of 8% (v/v) paraformaldehyde in PBS. Fix the cells with continuous gentle agitation for 15 min at room temperature.
3. Wash the fixed cells twice in 50 ml sterile PBS and resuspend the final pellet at 1×10^7 cells per ml in sterile PBS.
4. Prepare a 1:1 (v/v) emulsion of the cell suspension with Freund's complete adjuvant (FCA) and immediately inject 1 ml of the mixture intraperitoneally into mice of the same strain used to generate the idiotype monoclonal.

Protocol 1. *Continued*

5. Three weeks later boost the mice with cells prepared in the same way but in incomplete Freund's Incomplete adjuvant (FIA).

6. Test bleed the animals 8 to 10 days later. Assay the sera for inhibition in the functional assay. If positive, consider immediate myeloma fusion (see text).

7. Boost the mice with adjuvant-free fixed hybridoma cells at 6 weeks and repeat the test bleeds 8 to 10 days later.

8. If the anti-idiotype response persists, plan to use the mice for monoclonal antibody production. The final boost should consist of isolated monoclonal immunoglobulin administered intravenously in a dose of 20–100 μg 5 days before the fusion.

Table 1. Reagents and equipment for protocols

Reagents

Normal lymphocyte medium (NLM)	Alpha minimum essential medium (MEM); 20% (v/v) heat inactivated fetal calf serum (FCS); 10 mM Hepes, pH 7.2; 1 mM sodium pyruvate; 1 × Gibco non-essential amino acids; 50 μM 2-mercaptoethanol
Handling buffer (HB)	Alpha MEM; 10 mM Hepes, pH 7.2; 50 μM 2-mercaptoethanol
Thymocyte growth medium	HB with the addition of 2% (v/v) normal rabbit serum
Polyethylene glycol (PEG)	Reagent grade PEG; M_r = 20 000
Fusogen	PEG 1500, pre-packaged in 75 mM Hepes (Boehringer Mannheim no. 783 641)
Freund's complete adjuvant (FCA)	e.g. Sigma F5881
Freund's incomplete adjuvant (FIA)	e.g. Sigma F5506
Fusion plating medium	NLM (freshly prepared, supplemented with 2 mM glutamine if basal medium is more than 2 weeks old); 100 μM hypoxanthine (e.g. Sigma H9377)[a] 11.6 μM azaserine (2 μg/ml) (e.g. Sigma A4142)[a]. In the first feeding after a fusion the hypoxanthine and azaserine are added at twice this concentration to compensate for the volume of medium already in the wells

Equipment

- Sterile tissue culture plastic disposables
- Three pairs of small toothed forceps
- Two pairs of small straight dissection scissors
- 18G needles; 1-ml syringes
- Tissue culture facilities
- Dissection board
- Inverted phase contrast microscope
- Refrigerated bench centrifuge

[a] May be stored frozen as 25 × stock.

The anti-idiotype response may show an unusual or unexpected time course in comparison with the immune response to an antigen which is not a syngeneic immunoglobulin. For example, it is often observed that the initial response at 2–3 weeks after immunization gives the best anti-idiotype reactivity in the serum of the recipient animal (9). Subsequent boosts with idiotype immunoglobulin may result in the reduction or even disappearance of this response. Insufficient experience has yet accumulated to say whether or not this means that the animals should be taken for fusion as soon as they show an anti-idiotype response. In our hands, however, the fusion is normally carried out after a primary immunization and a single boost. The conventional fusion procedure given in *Protocol 4* may be used for the production of hybridomas after this type of immunization.

2.2 Generation of monoclonal anti-idiotype antibodies by paired *in vitro* immunization

In the absence of a functional assay for the selection of a single relevant idiotype, the best idiotype pool would consist of similar amounts of all antibodies against all of the epitopes present in the antigen, regardless of their relative immunogenicity. Such a mixture cannot normally be obtained from antisera because the mixture of idiotypes produced by repeated *in vivo* immunization is strongly shaped by the relative strengths of the epitopes and the contribution of T cell modulation. A closer approach to this ideal may be obtained from primary *in vitro* immunization, in which all possible B cell responses occur and the amplification due to the repeated stimulation of an immunodominant clone is not seen.

It would be possible to attempt to collect all of the idiotypes by myeloma fusion and hybridoma selection, but this introduces another level at which the response may become restricted. A technically simple and theoretically attractive alternative is to take the medium conditioned by the *in vitro* response, assume that it contains small amounts of all of the idiotype immunoglobulins secreted as a result of the initial immunization, and use this mixture as the antigen for a second round of *in vitro* immunization. In this way, all of the idiotypes are presented to the immune system for the generation of anti-idiotypes, which can then be rescued from the second *in vitro* immunization by myeloma fusion and hybridoma formation. Provided that the reconstituted immune systems immunized *in vitro* are genetically identical, no response to the constant region framework of the idiotype antibodies will be seen.

One way of speeding up the immunization required to produce the polyclonal response to the starting antigen and, subsequently, the polyclonal anti-idiotype response is to use *in vitro* immunization for both steps (10, 11). This approach has the advantages of speed and simplicity. It results in the production of monoclonal antibodies rather than highly variable polyclonal anti-idiotype antisera and also has the theoretical advantage that the immune response both to the starting antigen and subsequently to the idiotype mixture are as complete as possible because antigen-specific T suppression is circumvented (12). It is also possible that the predominance of pentavalent IgM in the idiotype mixture may enhance the anti-idiotype response. A potential disadvantage is that the second-round immunization also elicits a primary response, and the candidate monoclonal anti-idiotype antibodies will also be predominantly of the IgM isotype.

Protocol 2 describes the production of thymus-conditioned medium (TCM), which is a rich source of cytokines required to support the *in vitro* immune response. *Protocol 3* describes in detail the *in vitro* immunization technique itself, which is based upon the method described by Borrebaeck and colleagues (13), while *Protocol 4* describes a fusion procedure which has been tested for use with *in vitro* immunized spleen cells. *Protocol 5* outlines the use of paired *in vitro* immunizations for the production of monoclonal anti-idiotype antibodies.

Protocol 2. Preparation of thymus-conditioned medium (TCM)

1. Kill 3-week-old Balb/c and C57/B1 mice (not by cervical dislocation: ether, chloroform, and carbon dioxide are all acceptable inhalational agents). Final thymocyte yield should be 1–2 $\times$ 10^8 per mouse, and this means about 40 ml of TCM from a pair of mice (i.e. one of each strain). We usually prepare TCM using two mice of each strain.
2. Remove the thymus by dissection from the ventral surface. Ideally, the dissection should be performed in a laminar flow hood separate from that used for tissue culture. Pin the mouse on its back and swab with 70% ethanol. Open the skin and thoracic cavity in separate layers. The thymus is the bi-lobed midline white organ just above the heart. Remove it by careful dissection. Take care not to puncture the heart, trachea, or oesophagus as this greatly increases the risk of infection in the culture. The thymus is a substantial organ in 3-week-old mice, but it is not very robust; it may be necessary to lift out the two lobes separately.
3. Transfer the thymus to a 100-mm Petri dish containing 10 ml of handling buffer (HB) (see *Table 1*). Collect the Balb/c and C57/B1 thymuses into separate dishes.
4. In a tissue culture hood, dissect off fat and fibrous tissue and transfer the thymuses to new Petri dishes also containing 10 ml of HB (still keeping the strains separate). Tease the thymuses apart with two 18G needles until only small pieces of fibrous material remain. Alternatively, use a fine sterile stainless steel tissue sieve and force the thymus through with the rubber-tipped insert from a disposable syringe.

5. Transfer the HB containing the cell suspension to 50-ml conical tubes. Wash the Petri dishes with a further aliquot of HB and pool. Stand the tubes at room temperature for 5 min to allow any remaining large aggregates to settle.
6. Decant the supernatants without clumps into new tubes and centrifuge at 400 *g* for 10 min. Resuspend the cell pellets in HB and count using a haemocytometer or an electronic counter (e.g. Coulter counter).
7. Prepare suspensions of thymocytes from each mouse strain at a density of 5×10^6/ml in thymocyte growth medium (TGM; see *Table 1*). Mix 15 ml of each of these suspensions together in a 75 cm^2 tissue culture flask. Culture the flasks upright at 37 °C in a 5% CO_2 atmosphere. (The final density is 2.5×10^6/ml each of Balb/c and C57/B1 thymocytes in a total volume of 30 ml per flask). It is important to keep the flasks upright to maximize cell–cell contact and hence lymphokine production.
8. After 48 h examine the culture. The cell density should have increased visibly. Harvest the supernatant by centrifugation. Filter through a 0.22-μm sterile filter. The TCM may be used immediately or stored frozen at −20 °C. This preparation is usable for at least 2 months for the support of *in vitro* immunization, although we usually prepare it fresh. Frozen TCM remains an effective additive to support the cloning of delicate cells for at least 6 months.

Protocol 3. *In vitro* immunization

1. Kill a Balb/c mouse. Pin it out on its back and swab with 70% ethanol. Carefully open the abdominal skin only and retract. Using a fresh set of sterile instruments open the underlying translucent peritoneal layer separately.
2. The spleen is an elongated dark-red organ lying in the right upper quadrant (as you dissect from the ventral surface) of the peritoneal cavity. It is usually at least partly hidden by overlying stomach and small intestine. Carefully dissect out the spleen without opening the gut. Note that the spleens of non-immunized mice are neither as large nor as friable as the spleens of immunized animals. Transfer the spleen to a 100-mm Petri dish containing 10 ml of ice-cold HB and take this to a tissue culture hood.
3. In a tissue culture hood, dissect off any visible fat and transfer the spleen to a new Petri dish also containing 10 ml ice-cold HB. Tease the thymuses apart with two 18G needles until only small pieces of fibrous material remain. Alternatively, dice the spleen with a sterile blade and then use a fine sterile stainless steel tissue sieve and force the spleen fragments through with the rubber-tipped insert from a disposable

Protocol 3. *Continued*

syringe. Note that some methods used for the preparation of cells from a hyperimmunized spleen, such as washing a jet of buffer through the spleen using a needle and syringe, are not suitable. This is because such methods enrich for rapidly dividing cells from splenic foci low in connective tissue, which is ideal if the animal is already responding to the antigen of interest but unhelpful if, as in this case, the widest representation of an unimmunized spleen is required.

4. Transfer the suspension to a 50-ml conical tube. Wash the Petri dish with aliquots of the same ice-cold HB buffer and pool. Stand the tubes on ice for 5 min to allow large aggregates to settle.
5. Decant the supernatant free of large aggregates into a new tube and centrifuge at 400 *g* for 10 min. Resuspend the pellet in HB and count with a haemocytometer or Coulter counter.
6. Culture the unfractionated spleen cell suspension at 10^7 nucleated cells/ml in a medium consisting of 67% normal lymphocyte medium (NLM; *Table 1*) and 33% TCM final at 37 °C in a 5% CO_2 incubator. The final volume should be between 20 and 30 ml.
7. Add the sterile antigen to the culture in a small volume (up to 1.0 ml) of compatible buffer. A convenient way to ensure sterility of the antigen is to dilute it into the TCM and then refilter this mixture directly into the culture flask containing the cells in NLM through a 0.22-μm filter pre-wet with NLM to minimize protein absorption on the filter membrane. The antigen may be purified protein, synthetic peptide (which we normally use as a free peptide without conjugation to a carrier protein), or a complex mixture of proteins such as an organelle fraction. The required antigen dose is difficult to provide guidelines for; we aim to immunize with 1–10 μg of a purified protein, and normally use 10–200 μg of a synthetic peptide. For a complex mixture the possible presence of a few immunodominant antigens makes prediction impossible.
8. Leave the flask undisturbed at 37 °C in a 5% CO_2 incubator for 4 days. Ideally, place the flask flat so that it can be inspected without moving it; fungal or bacterial contamination is a risk with primary cultures and the telltale signs of cloudy acid medium are easily recognized. We do not routinely use antibiotics in the culture medium, preferring instead to know quickly if contamination has occurred.
9. On the fifth day examine the flask by inverted phase contrast microscopy. It should contain many dead cells, cell fragments, and clumps of red cells (see text and figures). In addition there should be a large population of adherent cells including macrophages and fibroblasts. There should also be a population of large well-spread adherent cells covered with grape-like clusters of spherical cells.

10. If all is well, harvest the non-adherent cells by briskly shaking the flask and pelleting all the released cells by centrifugation at 400 *g* for 5 min. Wash the flask with 30 ml of HB and centrifuge the wash separately. Keep the undiluted *in vitro* immunization supernatant. Pool the cells and wash twice in HB. This supernatant may be tested for the presence of antibodies reactive against the immunizing antigen, or used as the immunogen for a second round of *in vitro* immunization for the generation of monoclonal anti-idiotype antibodies.

11. The cells recovered from the *in vitro* immunization are used in place of the spleen input for a conventional myeloma fusion to generate hybridomas. If the pellet is more than slightly red it is helpful to lyse the erythrocytes at this stage. For hypotonic lysis, slowly add 5 ml of sterile ice-cold 0.2% saline to the pellet over 30 sec. *Immediately* add 5 ml of ice-cold 1.6% saline over 30 sec. Dilute the entire mixture rapidly into at least 30 ml of HB and centrifuge again (400 *g*, 5 min).

12. The spleen cells are now counted and used as the input for a splenocyte–myeloma fusion. Note that this protocol yields 1–2 $\times$ 10^7 cells, which is substantially less than the number of cells obtained from the spleen of a conventionally immunized animal. Many of the small lymphocytes originally placed in culture presumably do not survive under these culture conditions.

The appearance of a developing *in vitro* immunization culture is shown in the phase contrast photomicrographs (*Figures 1–5*). In *Figures 2–5* the upper panel shows an undisturbed culture, and illustrates the appearance of a healthy *in vitro* immunization at the intermediate stages before harvesting and fusion. The lower panel in these figures shows a culture which has been gently washed with warmed divalent cation-free PBS to reveal the underlying adherent cells and the cell clusters attached to them. Once washed in this way cultures are no longer suitable for further incubation, harvesting, and fushion because too many of the necessary cell–cell contacts have been disrupted. In this example a highly immunogenic preparation of viral spike glycoprotein has been used as the antigen; the features shown should also be present in cultures responding to weaker antigens, but may not be so prominent. Cultures established in the absence of antigen usually produce small numbers of clusters (much less than one per high power field at 320$\times$ magnification), which may represent a response to proteins of the fetal calf serum (FCS) in the medium, or simply the mitogenic effect of exposure to tissue culture plastic surfaces.

Figure 1 shows the culture within 10 min of setting up the *in vitro* immunization. The cells are at a uniform high density and almost completely mono-disperse. Many erythrocytes are visible, and a variety of larger cell

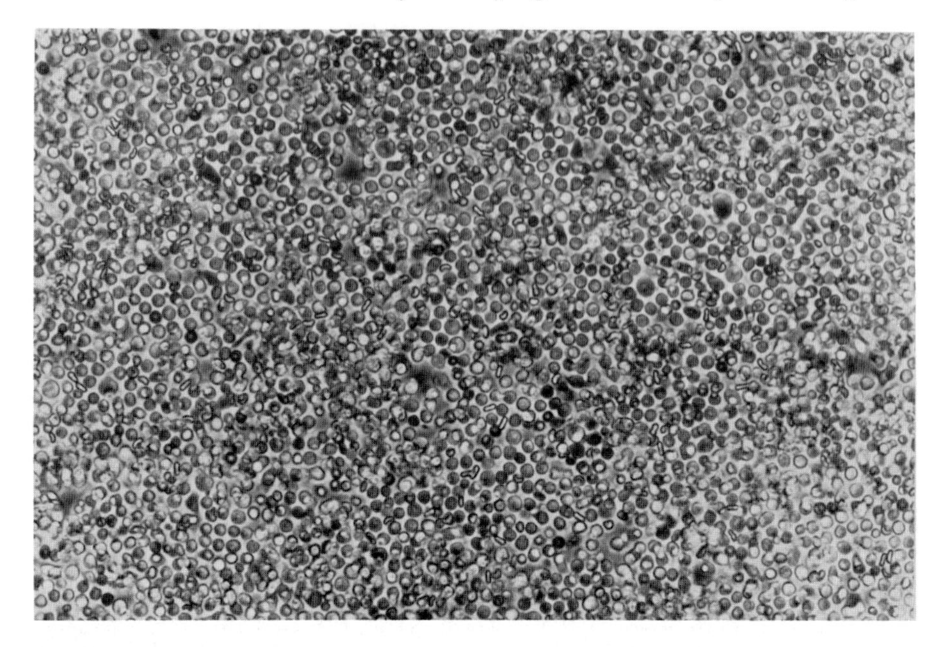

Figure 1. Phase contrast micrograph of an *in vitro* immunization culture immediately after setting up the culture. Final magnification, ×1092.

types which cannot be easily identified. No cells are adherent at this early stage.

Figure 2 shows the culture 24 h later. By this time inhomogeneities have begun to appear in the cell layer, and adherent cells are apparent (lower panel). Some of the adherent cells have small numbers of spherical high phase contrast cells attached to their surface. Substantial amounts of cell debris are visible even after washing; this material is cleared by macrophages in the culture. Debris-laden macrophages are widespread at 48 h, where they may be clearly seen as well spread flat cells with multiple pseudopodia tipped by active membrane ruffles, seen as dark margins by phase contrast (*Figure 3*: lower panel). Also at 48 h, clusters of large high phase contrast spherical cells become visible even in undisturbed cultures (*Figure 3*: upper panel, centre and top right of field). After washing, these developing clusters are seen to be attached to underlying adherent cells (*Figure 3*: lower panel, centre of field). We have made the empirical observation that *in vitro* immunizations that do not develop these clusters are unlikely to give rise to antigen-specific antibody-secreting hybridomas after fusion. In our experience the most common cause for this failure is a defect in the TCM or the FCS used in preparing the medium.

On day 4, 48 h later, the culture appears markedly inhomogeneous and the cell number has begun to drop (*Figure 4*: upper panel). The clusters of

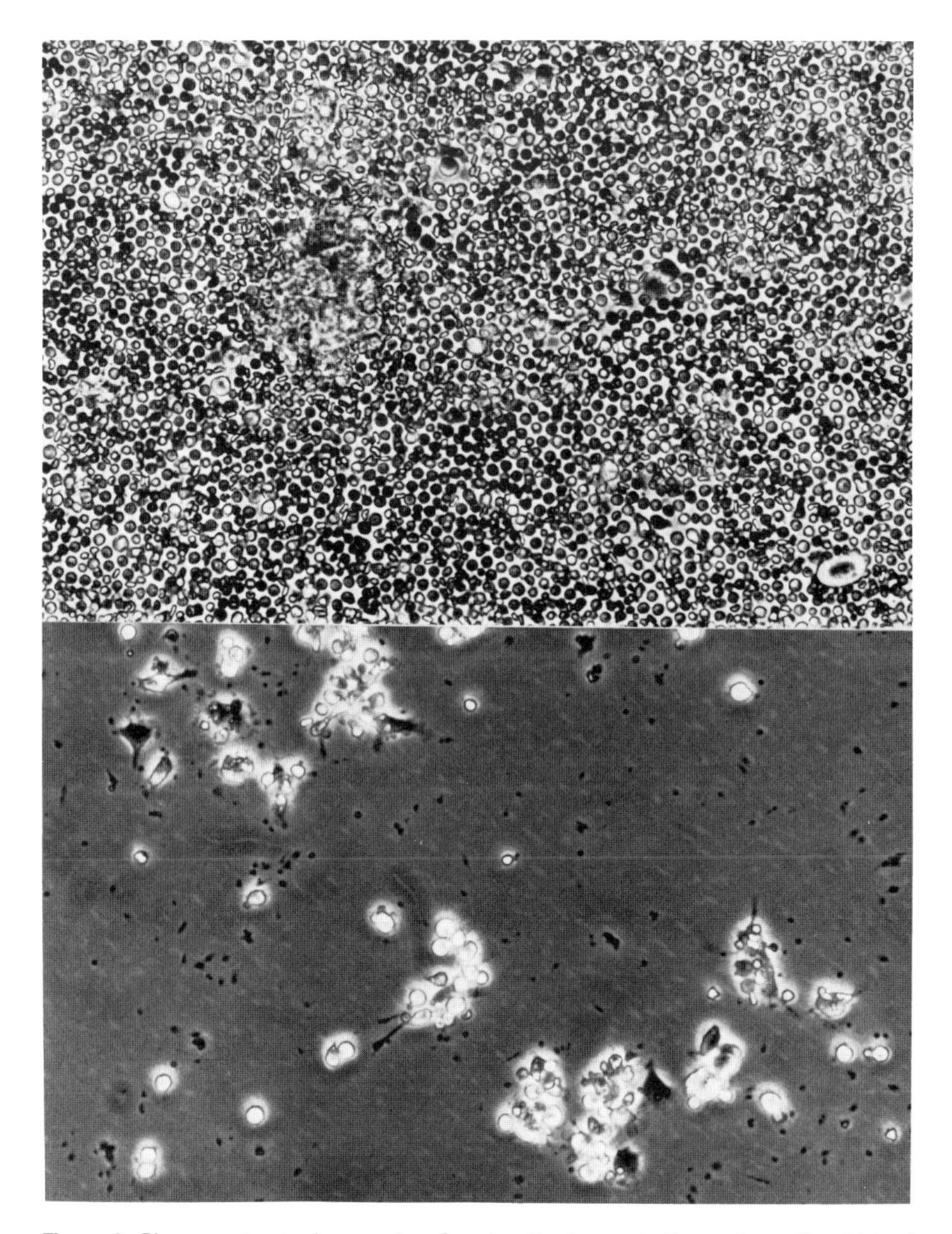

Figure 2. Phase contrast micrographs of an *in vitro* immunization culture after 24 h of culture. The upper panel shows the undisturbed culture and the lower panel shows the same culture after gentle washing with warmed divalent cation-free PBS to reveal the adherent cell population. Final magnification, ×1060.

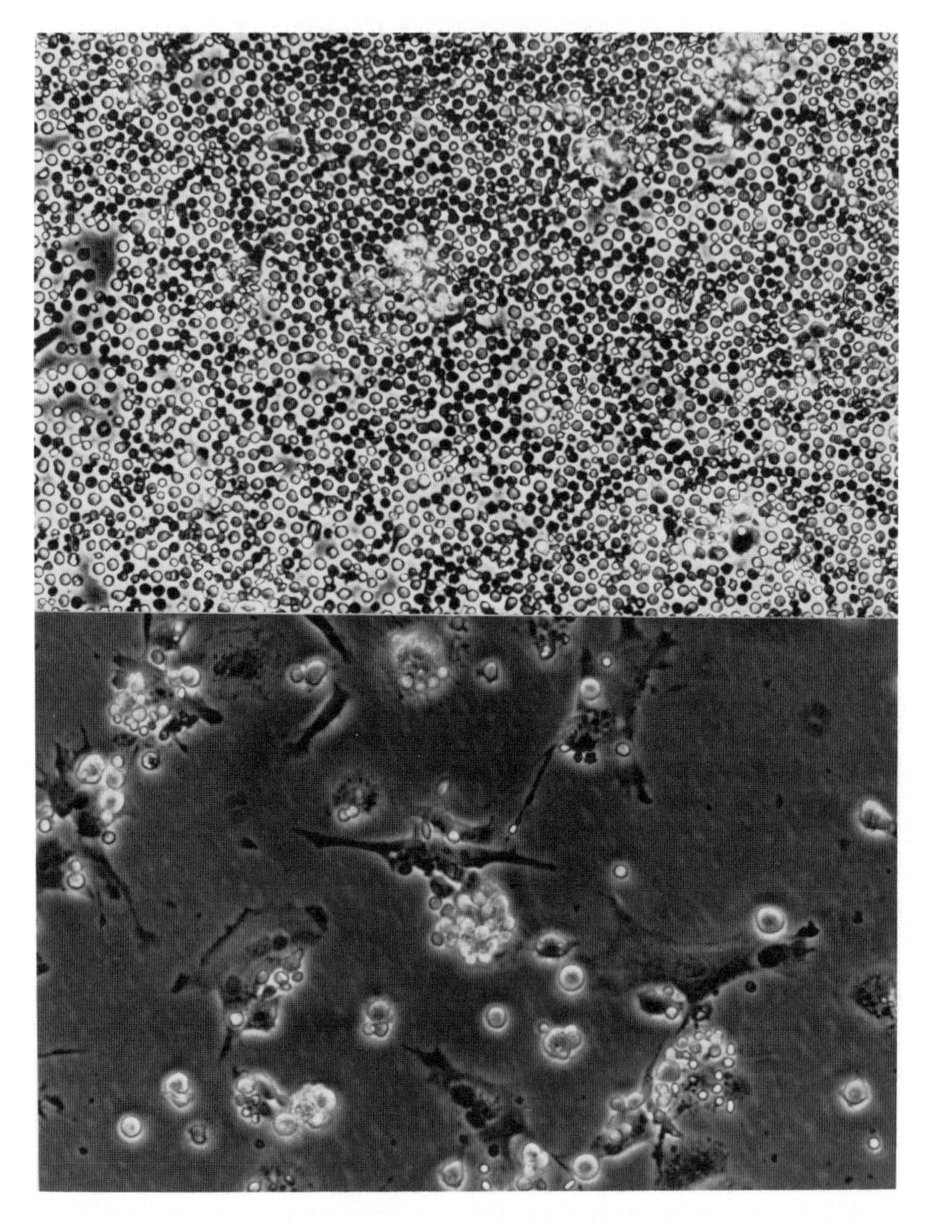

Figure 3. Phase contrast micrographs of an *in vitro* immunization culture after 48 h of culture. The upper panel shows the undisturbed culture and the lower panel shows the same culture after gentle washing with warmed divalent cation-free PBS to reveal the adherent cell population. Final magnification, ×1060.

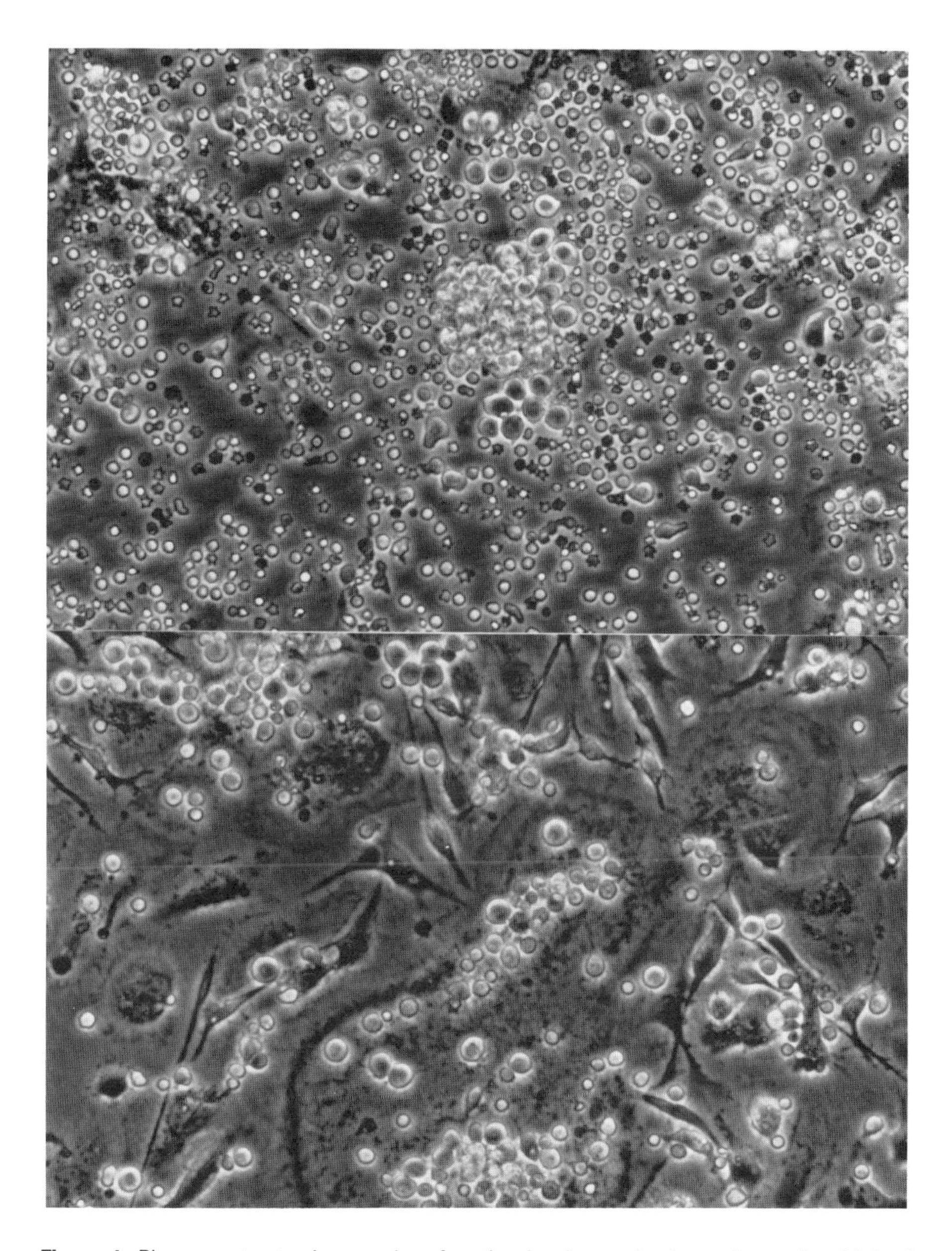

Figure 4. Phase contrast micrographs of an *in vitro* immunization culture after 96 h of culture. The upper panel shows the undisturbed culture and the lower panel shows the same culture after gentle washing with warmed divalent cation-free PBS to reveal the adherent cell population. Final magnification, ×1060.

refractile cells are now clearly visible. In a strong response clusters should be visible at a density of 2–3 per high power field (final magnification 320×). At this stage the culture should also contain clusters which are no longer attached to the plate, but are rolling freely in the medium. After washing, the 4-day culture is seen to contain a large heterogeneous population of adherent cells, including fibroblasts (visible as elongated bipolar cells), many macrophages (large well spread cells with intracellular debris and active membrane ruffles), and extremely large smooth-bordered flat cells (*Figure 4*: lower panel). This latter population often has substantial numbers of spherical refractile cells attached.

By day 5, the day of the fusion, the culture should look like the upper panel of *Figure 5*. Numerous large clusters of refractile cells are seen, most of which are no longer attached to the culture vessel. After washing to harvest the non-adherent cells for fusion, the residual cell population consists of fibroblasts, macrophages, and very large flat smooth-bordered cells which are frequently observed to be multinucleate (*Figure 5*: lower panel). Most of the refractile spherical cells have been released by the washing and will serve as input cells for fusion to the myeloma cell line to produce hybridomas.

Assuming that the *in vitro* immunization has an appearance similar to that of the upper panel of *Figure 4* or *5* on the fourth day, preparations are made for fusion on the following day. In our experience, the most common reasons for the absence of healthy clusters of proliferating cells in the *in vitro* immunization at this stage are:

- A bad batch of FCS. Most FCS sold as hybridoma tested is suitable, but occasionally poor results are seen. Repeating the immunization with a new batch usually cures the problem.
- Inactive TCM. While storage frozen for several weeks is probably acceptable, we now usually use freshly prepared TCM for *in vitro* immunizations.
- Toxic antigen. Concentrations of sodium dodecyl sulphate (SDS) above 0.1% in the antigen preparation may cause problems if the volume of antigen added is near the upper limit of 1 ml. The antigen should also be free of azide and organic solvents.
- Contamination with bacteria or fungi. Usually the result of contamination at the time of collecting the spleen, or due to the use of an inadequately sterilized antigen.

Protocol 4. Fusion after *in vitro* immunization

A. *During the week prior to the fusion*

1. Grow the chosen myeloma parent line (e.g. the SP-2/0-Ag14 line, available from the American Type Culture Collection as ATCC

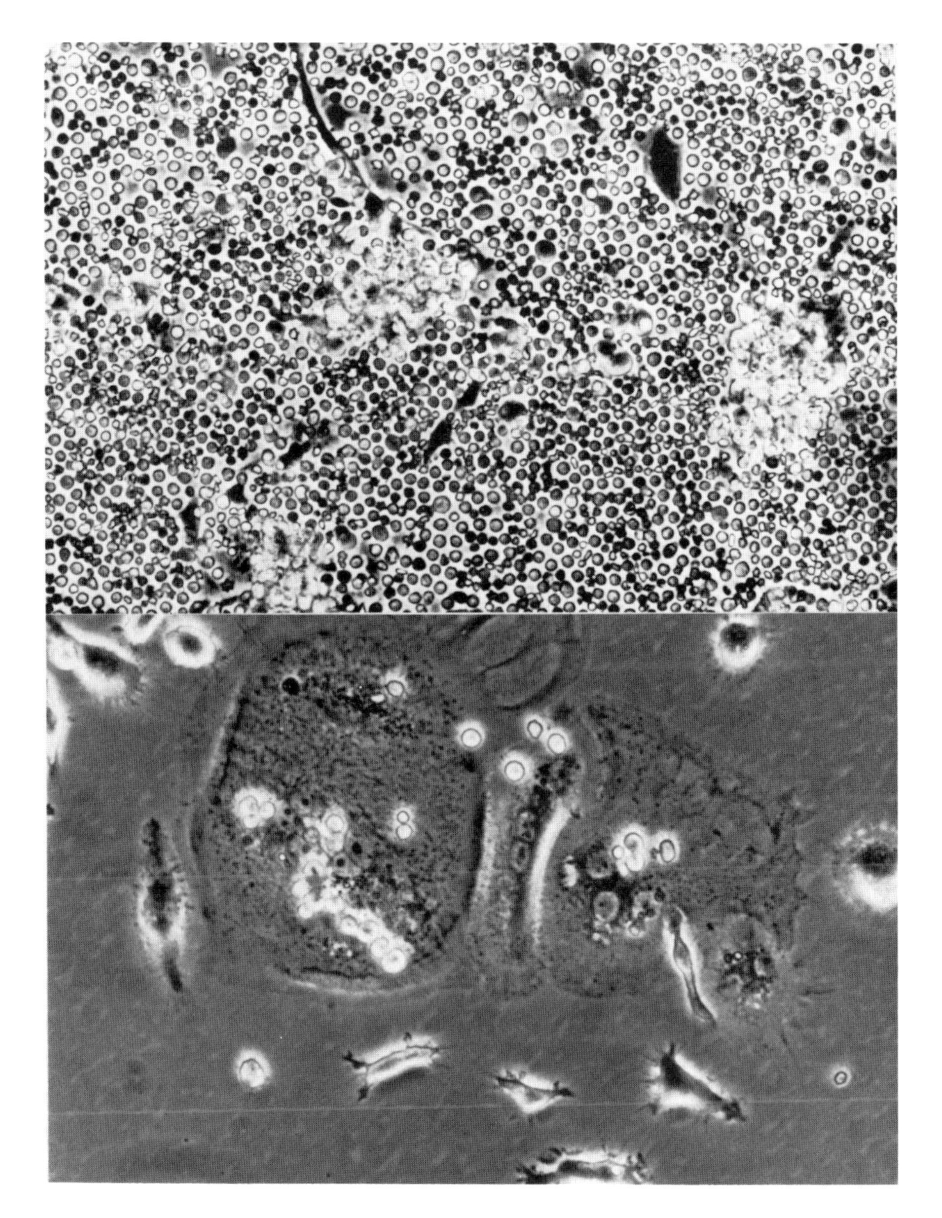

Figure 5. Phase contrast micrographs of an *in vitro* immunization culture immediately before harvesting the non-adherent and weakly adherent cells for fusion with myeloma cells to produce hybridomas (day 5; 120 h of culture). The upper panel shows the undisturbed culture and the lower panel shows the same culture after gentle washing with warmed divalent cation-free PBS to reveal the adherent cell population. Final magnification, ×1060.

Protocol 4. *Continued*

CRL1581) in NLM with very frequent feeding (i.e. subculturing at 3-fold dilution daily for the 4 days before the fusion).

2. Check the availability of all necessary reagents. Ensure adequate stocks of disposable tissue culture plastics.
3. Set up *in vitro* immunization (*Protocol 2*).
4. On the day before the fusion, check the *in vitro* immunization culture. Set up feeder plates consisting of 5 × 24-well plates containing macrophage feeders (approximately one Balb/c mouse per plate total peritoneal wash-out in 0.3 ml/well of NLM; do not wash off the non-adherent cells in the exudate as these seem to help).

B. *On the day of the fusion*

1. Harvest the non-adherent and weakly adherent cells from the *in vitro* immunization flask as described in *Protocol 2*.
2. Meanwhile, harvest the myeloma cells by shaking and centrifugation in the same way.
3. Wash both the pooled spleen cells and the myeloma cells twice in HB to remove all traces of serum. If the spleen cell pellet is uniformly red, perform the hypotonic lysis described in step 10 of *Protocol 2*. If the pellet only has a pink tinge or a red border then this lysis step is better omitted.
4. Separately count the myeloma cells and the nucleated cells in the *in vitro* immunized spleen cell pool using a haemocytometer or automated counter (e.g. Coulter counter). It may be difficult to count the relevant cells in the spleen cell suspension. The yield will be lower than from a spleen suspension prepared after *in vivo* immunization. It is possible to assume a yield of 2×10^7 spleen cells and fuse at a ratio of 2 spleen cells: 1 myeloma cell. Thus you need 10^7 myeloma cells and can plate the fusion out in five 24-well plates with a density of 8.3×10^4 myeloma input per well.
5. Mix all of the spleen cells with 10^7 myeloma cells and pellet together by centrifugation at 400 *g* for 5 min. Aspirate all but about 50 μl of the wash medium. Carefully resuspend the pellet in this last drop by gently tapping and flicking the tube. Try not to spread cells too far up the side of the tube.
6. Prepare conveniently to hand an opened vial of fusogen (Boehringer Mannheim PEG); *Table 1*, a tube containing 5 ml of HB, and a tube containing 10 ml NLM, together with a 1-ml, a 2-ml, and a 10-ml disposable sterile pipette. The fusion can be conveniently performed at

37 °C by holding the Falcon tube partly immersed in a beaker of warm water in the hood.

7. Add 1 ml of fusogen slowly and continuously to the cell suspension over 1 min with constant gentle shaking of the tube. Wait 1 min, shaking the tube gently every 10 sec. Add 2 ml of HB evenly over 2 min with continuous gentle shaking of the tube. Add 7 ml of NLM evenly over 3 min.
8. Centrifuge the fusion mixture at 300 *g* for 10 min. Aspirate the medium from the pellet carefully and resuspend very gently but thoroughly in 36 ml of fusion plating medium.
9. Plate out the fusion mixture at 0.3 ml/well in the prepared plates with macrophage feeders. It is important to be gentle as the fused cells are very fragile at this stage. Also plate out some myeloma cells at 1×10^5 cells/well in 0.3 ml NLM and 0.3 ml fusion plating medium on a separate plate as a control for the efficacy of the selection medium. These cells should be dead within 24 h.
10. After 24 h add 0.3 ml of fusion plating medium containing 2 × hypoxanthine and 2 × azaserine (see *Table 1*) to all the wells.
11. Feed the plates with 0.2 ml NLM on days 3, 5, and 7. Careful examination of the wells on day 7 should reveal small clusters of large round cells—these are what you have been waiting for. Screening can begin when there are 500–1000 hybridoma cells in the well. At a total magnification of 250× in the inverted phase-contrast microscope this is when there are the rough equivalent of two complete fields covered with hybridoma cells. The medium of wells containing this number of actively growing hybridoma cells is usually more yellow than that in surrounding unoccupied wells.

Protocol 5. Paired *in vitro* immunization for production of monoclonal anti-idiotype antibodies

1. Prepare TCM and set up an *in vitro* immunization as described in *Protocols 2* and *3*. Add sterile antigen containing the 'ligand'—e.g. synthetic peptide, isolated protein, or organelle fraction.
2. Check the *in vitro* immunization on the fourth day and prepare for myeloma fusion of the immunized cells if all is well.
3. On day 5 harvest the *in vitro* immunization culture as described and perform a myeloma fusion with the non-adherent cell population. Retain the supernatant from the culture (total volume should be approximately 30 ml).
4. Filter the supernatant through a 0.22-μm filter to remove

Protocol 5. *Continued*

cell fragments and dialyse the resulting filtrate against five changes of PBS (calcium and magnesium-free) overnight at 4 °C. Concentrate the filtered, dialysed supernatant 15-fold (i.e. to a final volume of approximately 2 ml) by dialysis against solid reagent grade polyethylene glycol (mean M_r 20 000; *Table 1*) at 4 °C.

5. The concentrated first-round supernatant may be used immediately as the antigen for the second round or stored frozen while the idiotype fusion is handled. If it is to be stored, make two 1-ml aliquots of the concentrate, flash-freeze them in liquid nitrogen, and store them at −80 °C.
6. Set up a second *in vitro* immunization culture using spleen cells of the strain used to generate the first round response. Add 1 ml of the first-round concentrate to 10 ml of TCM and 0.22-μm filter the mixture into the culture. If the concentrate has been frozen and thawed there may be a precipitate of FCS proteins which should be removed by centrifugation prior to filtration. Loss of mouse immunoglobulin at this step does not appear to be sufficient to prevent an anti-idiotype response.
7. Check the culture on day 4. There should be cellular proliferation and clusters of highly refractile cells, although these will probably be fewer in number than in the original immunization.
8. On day 5 harvest the cells and perform a standard myeloma fusion (see *Protocol 4*).
9. The yield of hybridomas from a second-round immunization is lower than that obtained from a single-round *in vitro* immunization and, if the fusion has been plated out as in *Protocol 4*, then 1–30% of the wells should contain hybrids.
10. The proportion of wells positive in the first assay is unpredictable because it is dependent both upon the immunogenicity of the ligand structure and the subsequent immunogenicity of the idiotype which is formed.

The utility of the paired *in vitro* immunization technique hinges on the availability of a screening assay for the desired internal image anti-idiotypes. Since the idiotype antigen is an unfractionated mixture, the relevant starting idiotype is not identified or separately isolated. It cannot therefore be used to test the putative anti-idiotype antibody by competition studies. There are two approaches to this problem. First, the cells from the first *in vitro* immunization may be subject to myeloma fusion to produce a panel of idiotype monoclonal antibodies, selected simply for their ability to recognize the starting antigen. If a member of this panel competes for antigen binding with the putative anti-idiotype antibody this identifies a pair of antibodies which merit further characterization. Alternatively, the putative anti-idiotype antibody may be selected by some indirect method based on assumptions about the protein–protein interaction under study and then used as antigen for a third round of immunization. If the putative anti-idiotype indeed bears an 'internal image' of the starting antigen, a subset of the third-round antibodies will not only bind the immunizing immunoglobulin but also recreate the specificity of the idiotype and bind to the starting antigen. If all three steps of such a network can be demonstrated then it is difficult to explain the process unless the intermediate anti-idiotype antibody is really an internal image (14). *Protocol 1* for production of anti-idiotype antisera from fixed hybridoma cells can be used for this step.

References

1. Jerne, N. K. (1974). *Ann. Immunol., Paris* **125C**, 373.
2. Burdette, S. and Schwarz, R. S. (1987). *New Engl. J. Med.* **317**, 219.
3. Coutinho, A. (1989). *Immunol. Rev.* **110**, 63.
4. Jerne, N. K. (1985). *EMBO J.* **4**, 847.
5. Bentley, G. A., Boulot, G., Riottot, M. M., and Poljak, R. J. (1990). *Nature* **348**, 254.
6. Sege, K. and Peterson, P. A. (1983). *Proc. nat. Acad. Sci., USA* **75**, 2443.
7. Langone, J. J. (ed.) (1989). *Methods in enzymology.* Vol. 178. Academic Press, San Diego.
8. Perelson, A. S. (1989). *Immunol. Rev.* **110**, 5.
9. Marriott, S. J., Roeder, D. J., and Consigli, R. A. (1987). *J. Virol.* **61**, 2747.
10. Reading, C. L. (1982). *J. immunol. Methods* **53**, 261.
11. Vaux, D. J. T. (1990). *Technique* **2**(2), 72.
12. Schrier, M. H. and Lefkovits, I. (1979). *Immunology* **36**, 743.
13. Borrebaeck, C. A. K. and Möller, S. A. (1986). *J. Immunol.* **136**, 3710.
14. Vaux, D. J. T., Helenius, A., and Mellman, I. (1988). *Nature* **336**, 36.

6

Protein targeting to mitochondria

ULLA WIENHUES, HANS KOLL, KARIN BECKER, BERNARD GUIARD, and FRANZ-ULRICH HARTL

1. Introduction

Over the last decade mitochondrial protein import has developed into one of the principal experimental systems for studying the general principles of intracellular protein sorting and the mechanisms involved in the translocation of proteins across membranes (for review, see references 1–5). Most mitochondrial proteins are synthesized as precursors on cytosolic polysomes and are subsequently imported into the pre-existing mitochondria which propagate by growth and division (*Figure 1*). The cytosolic precursors of mitochondrial proteins are maintained in a loosely folded, translocation-competent conformation by the interaction with chaperone components such as cytosolic hsp70. In general, precursors carry amino-terminal pre-sequences which contain necessary and sufficient information for the targeting to mitochondria. Mitochondrial targeting signals are typically positively charged and rich in hydroxylated amino acids. Precursors interact with receptor proteins of 19 and 72 kd (in *Neurospora crassa*) on the surface of mitochondria. For the insertion into the outer membrane most precursors appear to use a 'general insertion protein' (GIP) in the outer membrane. Proteins of 38 kd in *N. crassa* and 42 kd in *Saccharomyces cerevisiae* have recently been identified which may constitute at least part of GIP. Insertion at the level of GIP is dependent on ATP-hydrolysis which may be required for the release of factors bound to the precursor proteins. Translocation into and across the inner membrane probably occurs through some kind of proteinaceous channel or pore at so-called translocation contact sites where outer and inner membranes are held in close proximity. Precursors traverse contact sites in extended conformations. This translocation of the positively charged pre-sequences into or across the inner membrane is dependent on the electrical potential $\Delta\Psi$ across the inner membrane. Completion of translocation is aided by the interaction of the precursor chains reaching into the matrix with the mitochondrial hsp70 (ssc1 protein of *S. cerevisiae*). Mitochondrial hsp70 probably transfers the bound precursor in an ATP-dependent step to hsp60, a molecular chaperone in the matrix that mediates the folding

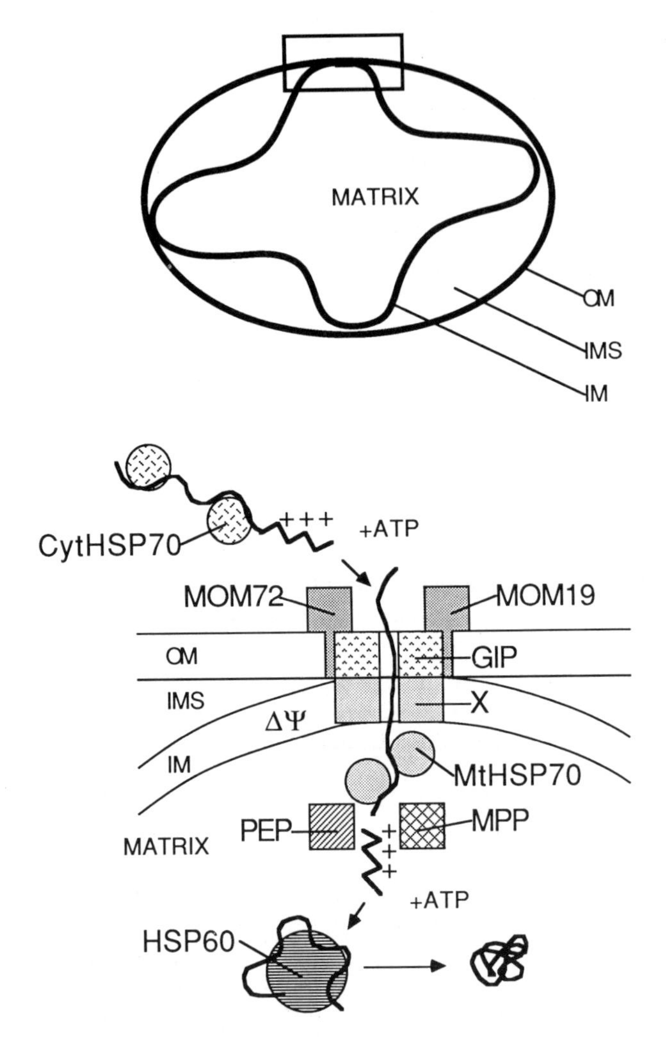

Figure 1. Working model for the import of precursor proteins. The import pathway of a precursor carrying a positively charged targeting sequence from the cytosol via contact sites into the matrix space of mitochondria is shown (see text for details). OM, outer membrane; IMS, intermembrane space; IM, inner membrane; $\Delta\Psi$, membrane potential across the inner membrane; CytHSP70, cytosolic hsp70; MtHSP70, mitochondrial hsp70; MOM72, mitochondrial outer membrane protein of 72 kd in *N. crassa* with receptor function; MOM19 mitochondrial outer membrane protein of 19 kd in *N. crassa* with receptor function; GIP, general insertion protein; X, unidentified component(s) of contact sites in the inner membrane; MPP, mitochondrial processing peptidase; PEP, processing enhancing protein.

and assembly of imported proteins in a further ATP-requiring reaction. Presequences are cleaved either during or after translocation by the metal-dependent processing enzyme in the matrix. Certain proteins of the inner membrane and intermembrane space then have to undergo further sorting steps within mitochondria in order to reach their target compartment.

The progress made in understanding the distinct steps of mitochondrial protein import and in identifying the components involved is largely based on the availability of an *in vitro* system allowing the faithful uptake of precursor proteins into isolated mitochondria. In the following sections we summarize a series of experimental procedures routinely carried out in our laboratory. Starting with the isolation of import-competent mitochondria and the synthesis of precursor proteins by *in vitro* translation or by overexpression in *Escherichia coli* we describe the assay system to study import of precursors into the isolated organelles. We then discuss the experimental approach for generating so-called translocation intermediates, i.e. precursor proteins caught in the act of translocation. The possibility of accumulating such intermediates on the transport pathway has been utilized in analysing the conformational restraints of protein translocation and in identifying membrane components functionally involved in the translocation process. We close by describing a set of methods for the subfractionation of mitochondria which enable the localization of further intermediates on the assembly pathway of proteins within mitochondria.

2. Preparation of mitochondria

The functional integrity of the organelles is a prerequisite for all studies of protein transport into isolated mitochondria. Intact mitochondria can be easily obtained from mammalian tissues which lack a rigid cell wall such as liver. However, mitochondrial protein import is best studied with fungal mitochrondria since these organelles, isolated from exponentially growing micro-organisms, exhibit high activities of protein import. Moreover, yeast has the advantage of being equally amenable to biochemical as well as genetic methods. We describe here two rapid procedures suitable for isolating milligram up to gram quantities of mitochondria from *N. crassa* or *S. cerevisiae* grown in liquid culture. Disruption of the cell wall is the critical step. The methods employed at that stage may interfere with the structural integrity of the organelles. In the case of *N. crassa* we have established a mechanical procedure, controlled grinding of the hyphae with quartz sand, whereas, with yeast, enzymatic digestion of the cell wall by zymolyase is prefered. The isolated mitochondria suspended in SEM buffer (see *Protocol 1*) can be stored in small aliquots at −80 °C after rapid freezing in liquid nitrogen. The frozen organelles preserve good activity in import experiments for weeks or months. They are thawed by a brief incubation at 25 °C.

Protocol 1. Isolation of mitochondria from *N. crassa*

Materials

- Vogel's minimal medium for growth of *N. crassa*: (8.4 mM Na_3 citrate × $2H_2O$, 36.7 mM KH_2PO_4, 25 mM NH_4NO_3, 0.8 mM $MgSO_4$, 0.68 mM $CaCl_2$, 1 nM biotin, 23.7 μM citric acid, 12.1 μM $Fe(NH_4)_2(SO_4)_2$, 17.3 μM $ZnSO_4$ × $7H_2O$, 1.5 μM $CuSO_4$, 0.2 μM $MnSO_4$, 0.8 μM H_3BO_3, 0.2 μM Na_2MoO_4
- 50% (w/v) sucrose
- SEM buffer: 250 mM sucrose; 2 mM EDTA; 10 mM MOPS/KOH, pH 7.2
- phenylmethylsulphonyl fluoride (PMSF); 200 mM in ethanol
- sterile quartz sand (Riedel de Haen)

A. *Preparation of Vogel's minimal medium supplemented with 2% (w/v) sucrose*

1. Make up 50× solution as follows
   - 150 g Na_3citrate × $2H_2O$
   - 250 g KH_2PO_4
   - 100 g NH_4NO_3
   - 10 g $MgSO_4$
   - 5 g $CaCl_2$

   Make up to 1000 ml with H_2O. This is solution 1.
2. Dissolve 5 mg biotin in 50 ml 50% ethanol and add 2.5 ml of this solution to solution 1. The resulting solution is solution 2.
3. Make up a 10 000× solutions as follows
   - 50 g citric acid
   - 10 g $Fe(NH_4)_2(SO_4)_2$
   - 50 g $ZnSO_4$ × $7H_2O$
   - 2.5 g $CuSO_4$
   - 0.5 g $MnSO_4$
   - 0.5 g H_3BO_3
   - 0.5 g Na_2MoO_4

   Make up to 1000 ml with H_2O and add 5 ml of this solution to solution 2. This is Vogel's minimal medium. (50× concentration).
4. Sterilize Vogel's minimal medium (50× concentration) by filtration.
5. Add 40 ml 50% (w/v) sucrose (autoclaved) to 20 ml Vogel's minimal medium (50× concentration) and make up to 1000 ml with autoclaved water.

B. *Isolation of mitochondria*

1. Inoculate the cell cultures[a] with 1 × 10^6 conidia/l Vogel's medium supplemented with 2% (w/v) sucrose.
2. Incubate *N. crassa* wild-type 74A (Fungal Genetic Stock Center N. 262) for 14 h at 25 °C under bright illumination with vigorous aeration. Introduce sterile compressed air via a glass tube.
3. Isolate hyphae from the cultures by suction filtration on to filter paper (Schleicher and Schüll, 595) using a Büchner funnel. Hyphae are peeled off the paper and are weighed. Stir the hyphae into a slurry with sterile quartz sand (1.5 g sand/g wet weight of hyphae) and SEM buffer (2 ml/g wet weight of hyphae). Add PMSF to a final concentration of 2 mM from a freshly prepared 200 mM solution in ethanol. Grind hyphae for 1–2 min with a mortar and pestle at 4 °C.
4. Wash the resulting slurry twice by centrifugation at 4 °C for 5 min at 2000 *g* (e.g. Beckman JA20 rotor) and discard the pellets (sand and cell walls).
5. Centrifuge the combined supernatants (cell homogenate) for 12 min at 17 400 *g* to obtain the mitochondrial pellet.
6. Wash mitochondria by resuspension in SEM + PMSF and centrifugation at 4 °C for 12 min at 17 400 *g*.
7. Resuspend mitochondria in SEM and adjust to 5 mg protein/ml based on Bradford protein assay (7).

[a] The cultivation of *N. crassa* has been described in detail (6).

Protocol 2. Isolation of mitochondria from yeast

Materials

- growth medium (YEPG): 2% bactotryptone, 1% yeast extract, 2% ethanol, 3% glycerol, adjusted to pH 5.0 with HCl
- Zymolyase 20T (Seikagaku Kogyo Co. Ltd)
- Tris/dithiothreitol (Tris/DTT): 0.1 M Tris/H_2SO_4, 10 mM DTT, pH 9.4
- sorbitol/KH_2PO_4: 1.2 M sorbitol/20 mM KH_2PO_4, pH 7.4
- homogenizing buffer: 0.6 M sorbitol, 10 mM Tris-HCl, pH 7.4, 1 mM EDTA, 0.1% fatty acid-free bovine serum albumin (BSA) (Sigma)
- PMSF: 200 mM in ethanol

Method

1. Dilute an overnight yeast culture[a] to optical density at 600 nm (OD_{600}) = 0.1.

Protocol 2. *Continued*

2. Grow cells under vigorous shaking (160 r.p.m.) for 15 h at 30 °C in YEPG medium to OD_{600} = 2–3.
3. Pellet cells by centrifugation for 5 min at 3000 *g* at room temperature.
4. Wash the cells by resuspension in distilled H_2O and centrifugation for 5 min at 3000 *g* at room temperature.
5. Weigh the resulting cell pellet and resuspend it in Tris/DTT (1 ml buffer/ 0.5 g wet weight of cells) using a glass pipette.
6. Incubate the cell suspension in an Erlenmeyer flask for 10 min at 30 °C in a shaking water bath (90 r.p.m.).
7. Pellet the cells at 4 °C by centrifugation at 3000 *g* for 10 min.
8. Wash the cells by resuspension in 1.2 M sorbitol and centrifugation for 10 min at 3000 *g* at 4 °C.
9. Resuspend the cells in sorbitol/KH_2PO_4 using 1 ml buffer/0.15 *g* wet weight of cells.
10. Add 1–2 mg Zymolyase-20T from *Arthrobacter luteus* per g wet weight of cells and incubate for 30–90 min at 30 °C in a shaking water bath (90 r.p.m.).[b]
11. Pellet the spheroplasts (5 min, 3000 *g*, 4 °C).
12. Wash the spheroplasts twice by resuspension in cold 1.2 M sorbitol + PMSF and centrifugation.[c]
13. Resuspend the spheroplasts in cold homogenizing buffer + PMSF (1 ml buffer/0.15 *g* wet weight of cells).
14. Spheroplasts are broken by 10–15 strokes in a tight fitting Dounce homogenizer cooled in ice water.
15. Mix the resulting homogenate with 1 volume of homogenizing buffer + PMSF and stir in an Erlenmeyer flask for 10–15 min at 4 °C to break remaining spheroplasts.
16. Centrifuge the resulting homogenate at 4 °C for 5 min at 2000 *g* and for 5 min at 3000 *g*. Discard the pellets (cell debris).
17. Pellet the mitochondria by centrifugation at 4 °C for 10 min at 15 000 *g*.
18. Wash the mitochondria twice by resuspension in SEM buffer + PMSF (*Protocol 1*) and centrifugation as in step **17**.

[a] Cultivation of yeast has been described in detail (8).
[b] Conversion of intact yeast cells to spheroplasts is controlled by measuring OD_{600} after 1:100 dilution in H_2O; the suspension is clear when all cells have formed spheroplasts. Shorter incubation in the presence of zymolyase usually yields fewer but more intact organelles.
[c] PMSF is added in this and all subsequent steps to a final concentration of 2 mM from a freshly prepared stock solution in ethanol to reduce protease digestion of mitochondrial proteins of the outer membane, e.g. import receptors.

3. Import of precursor proteins synthesized in reticulocyte lysate

3.1 Synthesis of radiolabelled precursor proteins

Various translation systems allowing the *in vitro* synthesis of proteins have been developed and most of them are commercially available, e.g. lysates of rabbit reticulocytes (9, 10), of wheat germ cells (11, 12), of yeast (13, 14), and of *E. coli* (15, 16). Radiolabelled precursors of mitochondrial proteins are efficiently synthesized in lysates of rabbit reticulocytes. This system has the advantge of containing the full complement of cytosolic factors that support the import of proteins into mitochondria. The synthesis reaction is performed in a two-step process involving transcription of DNA coding for the protein of interest followed by translation of the resulting mRNA in the reticulocyte lysate.

In most cases, cDNAs (or genomic DNA in the case of yeast) coding for the mitochondrial precursor proteins are cloned in pGEM4 plasmids (Promega Biotec) or Bluescript M13 plasmids (Stratagene) prior to use in the coupled transcription/translation reaction. The recombinant template-DNA's are linearized by cutting the 3′ non-coding region of the gene encoding the precursor with a suitable restriction enzyme. It is of advantage to use restriction enzymes that generate 5′ protruding ends or blunt ends (this favours the release of the RNA-polymerase following transcription). The choice of RNA-polymerase is dependent on the plasmid used for cloning and, with many vectors, also on the orientation of the insert in the plasmid. The most commonly used RNA-polymerases are derived from the bacteriophages Sp6, T3, and T7 (17, 18). The RNA-polymerase reaction is run at 37–40 °C. After transcription the polymerase is removed from the nucleic acids (mRNA and DNA) in the transcription mixture by extraction with phenol/chloroform/isoamylalcohol and the nucleic acids are ethanol-precipitated. The mRNA is dissolved in H_2O and added to the translation mixture containing rabbit reticulocyte lysate, radiolabelled methionine, a mixture of unlabelled amino acids, and optimal concentrations of magnesium and potassium (roughly 1–3 mM and 100–200 mM, respectively). Following translation, unlabelled methionine is added to compete for the labelled methionine and sucrose is added to a final concentration of 250 mM to achieve isoosmotic conditions. All solutions, reaction tubes, pipette tips, etc. used in the preparation of the template DNA, the transcription, and the translation reactions must be sterilized either by autoclaving at 120 °C or by sterile filtration (in the case where heating destroys the compound). Disposable gloves should be worn throughout the procedure.

Protocol 3. Transcription of the cDNA

Materials

- premix: 6 mM MgOAc; 2 mM spermidine; 10 mM DTT; 0.5 mM each of ATP, CTP and UTP; 0.1 mM GTP; 0.1 mg/ml BSA; 40 mM Hepes–KOH, pH 7.4
- capped guanosine (^{7}mG(5′)pppG) (Pharmacia): dissolved at a concentration of 1 mM in sterile 10 mM Hepes–KOH; solution is then sterilized again by filtration
- RNasin (40 U/μl)
- Sp6-, T3-, or T7-RNA-polymerase
- phenol/chloroform/isoamylalcohol: 50% (v/v) phenol equilibrated with 50 mM Tris-HCl, pH 8.0; 46% (v/v) chloroform; 4% (v/v) isoamylalcohol
- 10 M LiCl
- ethanol (100%)
- ethanol (70%)
- sterile water
- sterile pipette tips
- sterile reaction tubes

A. *Preparation of premix*

1. Mix all components (see *Materials*) from 10–100 × concentrated stock solutions prepared with autoclaved water.
2. Sterilize the complete premix again by filtration through a 0.2-μm filter. Do not autoclave the stock solutions of DTT, nucleotides, BSA, and spermidine.
3. Divide the final solution into aliquots, freeze in liquid nitrogen, and store at −80 °C. Stock solutions of the nucleotides are prepared in 10 mM Hepes–KOH, pH 7.4 and stored at −80 °C.

B. *Transcription*

1. Prepare a 200 μl transcription reaction consisting of:
   - 120 μl premix
   - 14 μl ^{7}mG(5′)pppG (1 mM)
   - 2 μl RNasin (40 U/μl)
   - 10 μg linearized plasmid-DNA (in H_2O)
   - 50 U of Sp6-, T3-, or T7-RNA-polymerase
   - sterile H_2O to a final volume of 200 μl

Add the plasmid-DNA and the polymerase last after warming the reaction mixture to room temperature.

2. Incubate for 1 h at 37–40 °C.
3. Add an equal volume (200 µl) of phenol/chloroform/isoamylalcohol, mix by shaking for 5 min, centrifuge for 5 min in an Eppendorf microcentrifuge. Remove the upper aqueous phase (discard the lower phenol/chloroform phase).
4. Add 200 µl chloroform to the aqueous phase, mix vigorously for 2 min; centrifuge again for 5 min and collect the upper aqueous phase (discard the lower chloroform phase).
5. Add 20 µl LiCl (10 M) and 600 µl ethanol (100%) to the aqueous phase, mix thoroughly, and incubate for 1 h at −20 °C.
6. Centrifuge at 20 000 *g* (e.g. Beckman JA-20 rotor equipped with adaptors for microfuge tubes) for 25 min at 2 °C, carefully remove the supernatant, and wash the pellet by adding 500 µl ethanol (70%) without mixing.[a]
7. Centrifuge briefly at 20 000 *g* at 2 °C; again remove the supernatant and dry the pellet under vacuum in a desiccator until all liquid has disappeared.
8. Dissolve the mRNA-pellet carefully in 200 µl H_2O (add 0.1 U RNasin/µl). Divide the solution into 50 µl aliquots and store at −80 °C.

[a] Washing the RNA pellet with 70% ethanol is not absolutely required. Li^+ ions do not disturb the translation reaction up to a concentration of 1 M; when using other salts in the precipitation of the nucleic acids (step **5**) washing is essential. Do not dry the RNA pellet completely (step **7**). It is then difficult to dissolve the RNA.

Protocol 4. Translation of the mRNA

Materials

- rabbit reticulocyte lysate (Amersham)
- 1000 Ci/mmol ^{35}S-labelled methionine
- amino acid mix excluding methionine
- 40 U/µl RNasin
- 0.5 M KOAc
- 20 mM MgOAc

A. *Preparation of amino acid mix*

1. Dissolve 19 amino acids excluding methionine at 2 mM each in sterile 10 mM DTT.

Protocol 4. *Continued*

2. Adjust to pH 7.5 with KOH and sterilize by filtration.[a]

B. *Translation reaction*

1. Prepare a 200 μl translation reaction consisting of:
   - 100 μl rabbit reticulocyte lysate[b]
   - 3 μl amino acid mix (minus methionine)
   - 0.5 μl RNasin (40 U/μl)
   - 20 μl [^{35}S]-methionine (1000 Ci/mmol)
   - variable volume of 0.5 M KOAc[b]
   - variable volume of 20 mM MgOAc[b]
   - 50 μl mRNA solution[b]
   - sterile H_2O to a final volume of 200 μl
2. Mix gently but thoroughly and incubate for 1 h at 30 °C.
3. Cool to 0 °C; add 5 μl of unlabelled methionine (0.2 M) (final concentration 5 mM) and 40 μl sucrose (1.5 M) (final concentration 250 mM).
4. Centrifuge at 150 000 *g* (Beckman Ti50 rotor with adaptors for microfuge tubes) for 20 min at 2 °C (removal of ribosomes).
5. Collect the supernatant (discard the pellet); divide the supernatant into 50-μl aliquots, freeze in liquid nitrogen, and store at −80 °C.

[a] Certain amino acids do not dissolve until pH is adjusted.

[b] Several parameters should be optimized for the synthesis of different proteins. (i) The amount of mRNA is titrated by adding increasing volumes of the RNA solution obtained in the transcription reaction. (ii) The optimum for the final concentration of Mg^{2+} ions is in the range of 1–3 mM; the optimal concentration of K^+ varies between 100 and 200 mM; in commercially available rabbit reticulocyte lysate the concentrations of Mg^{2+} and K^+ are specified and have to be taken into consideration. (iii) The portion of reticulocyte lysate in the final translation reaction has to be titrated.

3.2 Import of precursor proteins into isolated mitochondria

The import of mitochondrial precursor proteins into isolated mitochondria of yeast of *N. crassa* is carried out in MOPS buffer containing sucrose, BSA, K^+ and Mg^{2+} ions, DTT, respiratory chain substrates, and ATP. Fatty-acid-free BSA was found to stabilize mitochondria probably by binding free fatty acids. NADH or potassium ascorbate plus *N*,*N*,*N*′,*N*′-tetramethylphenylenediamine (TMPD) are added to establish a membrane potential across the inner membrane. The import of certain precursor proteins is highly dependent on the presence of cytosolic factors which are still poorly characterized. The concentration of cytosolic factors can be increased by using higher amounts of reticulocyte lysate in the import reaction (e.g. the import of the β subunit of F_1-ATPase requires a final concentration of up to 30% (v/v) reticulocyte lysate). Usually 5–10% (v/v) reticulocyte lysate in the import assay is optimal (19). For the import of many precursors the ATP contained in the added reticulocyte lysate is sufficient. The dependence of the import on the membrane potential across the inner membrane (except for outer membrane and some intermembrane space precursors) is taken as a criterion that faithful import has occurred (2, 20, 21). This dependence is tested by adding potential dissipating reagents to the import reaction such as the potassium ionophore valinomycin (1 μM final concentration in the presence of > 40 mM K^+), carbonyl cyanide *m*-chlorophenylhydrazone (CCCP; 10–50 μM final concentration; when using CCCP the DTT should be omitted) or antimycin A/oligomycin (8 μM/20 μM, final concentrations; only effective in mitochondria of *N. crassa*). Ascorbate/TMPD allows the re-establishment of a membrane potential in mitochondria of *N. crassa* whose inner membrane had been de-energized by combination of antimycin A and oligomycin, inhibitors of complex III and the FoF_1-ATPase, respectively. Ascorbate/TMPD feeds electrons into complex IV and thereby circumvents the block by antimycin A (20). The sequestration of the precursor by the mitochondria during import is tested by incubation of the organelles in the presence of externally added protease. Precursor unspecifically associated with the surface of the outer mitochondrial membrane is digested whereas the completely imported protein is resistant to protease treatment.

Protocol 5. Import of precursor proteins synthesized in reticulocyte lysate

Materials

- import buffer: 250 mM sucrose; 3% BSA; 10 mM MOPS, pH 7.2; 80 mM KCl; 5 mM $MgCl_2$; 5 mM unlabelled methionine
- isolated mitochondria of yeast or of *N. crassa* resuspended in SEM buffer at 5–10 mg/ml. 0.2 M NADH in SM buffer: 250 mM sucrose; 10 mM MOPS, pH 7.4

Method

1. Prepare a 200 μl import reaction containing import buffer, 2 mM NADH, 50 μg mitochondria, and 1–30% (v/v) reticulocyte lysate containing radiolabelled precursor (see Section 3.1.2). Incubate for 20 min at 25 °C.[a]
2. Cool to 0 °C on ice.
3. Divide the reaction into halves.
4. Add proteinase K to one-half (15–200 μg/ml final concentration) and incubate for 10–30 min at 0 °C.[b]

Protocol 5. *Continued*

5. Stop protease action by addition of 1 mM PMSF and incubate for 5 min at 0 °C; PMSF is also added to control samples minus protease.
6. Re-isolate mitochondria by centrifugation for 10 min at 15 000 *g* at 2 °C (Beckman JA-20 rotor equipped with adaptors for microfuge tubes).
7. Resuspend mitochondria in suitable buffer for further experimentation or wash the mitochondrial pellet in 300 μl SEM buffer (without resuspending) and solubilize in SDS containing buffer after a brief (3 min) centrifugation at 15 000 *g* for analysis by SDS-PAGE and fluorography.

[a] Shorter incubation may be sufficient for complete import of the precursor added. The structural integrity of the mitochondria may suffer somewhat in the presence of reticulocyte lysate. With various precursor proteins the optimal concentration of KCl is in the range of 40–150 mM. The application of other salts (e.g. sodium or potassium phosphate) may increase the efficiency of import. The optimal concentration of added MgOAc is 0–5 mM; addition of 0.5–2 mM ATP may be necessary if the proportion of reticulocyte lysate in the import assay is low.

[b] Conditions for protease treatment may vary for different precursor proteins. Protease treatment should result in complete digestion of the precursor protein in control reactions without mitochondria or in reactions containing micochondria whose membrane barrier has been broken by detergent or by sonication (20).

4. Import of precursor proteins expressed in *E. coli*

4.1 Expression and purification of precursor proteins

For the study of certain kinetic and mechanistic aspects of protein translocation into mitochondria purified precursor proteins available in large amounts are a prerequisite. Over recent years several mitochondrial precursor proteins have been successfully expressed in *E. coli* and have been purified from inclusion bodies in large quantities (22, 23). To allow high level expression of mitochondrial precursors in *E. coli* the corresponding genes are cloned under the control of efficient prokaryotic promoter regions in expression vectors such as pJLA and pUHE (24, 25). Two types of promoter–operator sequences are particularly useful. Both are tightly repressed in the absence of induction.

- Lac promoters are tightly repressed in *E. coli* strains that produce the lac represser in large quantities, but highly active upon induction with isopropylthiogalactoside (IPTG).
- Thermosensitive phage λ promoters are repressed at reduced growth temperature (28 °C), but are active at high growth temperature (41 °C).

The level of gene expression may vary with the individual *E. coli* strain used and is dependent on certain structural features of the DNA constructs. In general, the gene has to be cloned in an optimal relationship to efficient transcriptional and translation initiation signals of procaryotic origin. Furthermore, the level of gene expression depends on the 5′ sequence of the individual gene cloned for expression; *in vitro* mutagenesis of the DNA resulting in conservative exchanges of single amino acid residues may provide more favourable structures. Heterologous gene expression in *E. coli* has been extensively reviewed (26–8).

We describe in this section the expression of an artificial mitochondrial precursor, a fusion protein containing the first 167 residues of precytochrome b_2 and the complete sequence of mouse dihydrofolate reductase (DHFR). Cloning was carried out in a pUHE vector as well as in a pJLA vector. Similar fusion proteins have been useful in establishing the targeting signals of mitochondrial proteins and in elucidating distinct steps of mitochondrial protein import including the unfolding and refolding of precursor proteins during import (for review see reference 29).

Protocol 6. Expression using an IPTG-dependent promoter

1. Start an overnight culture in L-broth (LB) medium with a single colony of the appropriate strain of *E. coli* using the required selection conditions. Incubate at 37 °C with vigorous shaking (160 r.p.m.).
2. Dilute the overnight culute 1:1000 in fresh LB medium containing selective agent.
3. Grow the cells for 2–3 h at 37 °C with shaking to an OD_{585} = 0.5.
4. Induce expression by adding IPTG to 1 mM (from a 100 mM stock solution).
5. Continue growth for 1 h.
6. Harvest the cells by centrifugation at 12 000 *g* for 10 min.

Protocol 7. Expression using a thermoinducible promoter

1. Start an overnight culture as in *Protocol 6*, step **1**. Incubate at 28 °C with vigorous shaking.
2. Dilute the overnight culture 1:1000 in fresh LB medium containing selective agent.
3. Grow for 2–3 h to an OD_{585} = 1.0 with shaking.
4. Induce expression by diluting the culture with an equal volume of medium pre-warmed to 54 °C to rapidly reach the induction temperature of 41 °C.
5. Grow the cells with shaking for 1 h at 41 °C.
6. Harvest the cells by centrifugation at 12 000 *g* for 10 min.

Protocol 8. Radiolabelling of overexpressed precursor proteins

A. *Preparation of low-sulphate medium*

Materials

Final concentrations in low-sulphate medium

- 40 mM MES–KOH, pH 7.4
- Tricine–KOH, pH 7.4
- 10 μM $FeCl_3$
- 10 mM NH_4Cl
- 250 μM KCl
- 50 mM NaCl
- 5 mM $MgCl_2$
- 5 μM $CaCl_2$
- 300 nM Mo_7O_{24}
- 30 μM boric acid
- 3 μM $CoCl_2$
- 1 μM $CuSO_4$
- 8 μM $MnCl_2$
- 2 μM $ZnCl_2$
- 1 mM KH_2PO_4, pH 7.4
- 40 μg/ml each of 18 amino acids excluding cysteine and methionine (added from 1 mg/ml stock solutions in 10 mM DTT, adjusted to pH 7.5 with KOH)
- selective antibiotic, e.g. ampicillin
- other substrates required by the individual *E. coli* strain

Method

1. Prepare a 10 × concentrated stock solution of 400 mM MES, 40 mM Tricine, 100 μM $FeCl_3$, 100 mM NH_4Cl, 2.5 mM KCl, 500 mM NaCl, 50 mM $MgCl_2$, 50 μM $CaCl_2$, and adjust to pH 7.4 with KOH.
2. Autoclave for 10 min at 120 °C.
3. Prepare a micronutrient mix containing 1000 × concentrated Mo_7O_{24} (300 μM), boric acid (30 mM), $CoCl_2$ (3 mM), $CuSO_4$ (1 mM), $MnCl_2$ (8 mM), and $ZnCl_2$ (2 mM), and sterilize by filtration through a 0.2-μm filter.
4. Amino acids are dissolved as a separate 25 × concentrated mix (1 mg/ml) which is sterilized by filtration. Concentrated solutions of vitamins and

antibiotics are also sterile-filtered. The sterilized stock solutions are stored at 4 °C. The final medium is prepared immediately before use.

B. *Radiolabelling*

1. Start an overnight culture in LB medium as in *Protocol 6*, step **1**.
2. Dilute the overnight culture 1:100 with low sulphate medium containing selective agent.
3. At the time of induction add ^{35}S-labelled sulphate (3000 Ci/mol; Amersham to a final activity of 0.1 mCi/ml.
4. Continue as in *Protocol 6*, steps **3–6** or *Protocol 7*, steps **3–6**, depending on the method of induction.

Protocol 9. Extraction of inclusion bodies

Materials

- solution 1: 25% sucrose; 50 mM Tris-HCl, pH 8.0
- solution 2: 20 mM Tris-HCl, pH 7.4; 1 mM EDTA; 1 mM PMSF; 1% Triton X-100; 50 mM DTT
- solution 3: 20 mM Tris-HCl, pH 7.4; 1 mM EDTA; 1 mM PMSF; 0.1% Triton X-100; 50 mM DTT
- solution 4: 20 mM Tris-HCl, pH 7.4; 1 mM EDTA; 1 mM PMSF; 50 mM DTT
- solution 5: 7 M urea or 6 M guanidinium-Cl (SIGMA); 50 mM Tris-HCl, pH 7.4; 50 mM DTT

Method

1. Resuspend the pelleted cells (see *Protocols 6–8*) in solution 1 (2 ml/100 ml original cell culture). Add lysozyme (0.2 ml/100 ml original cell culture) from a 10 mg/ml stock in solution 1.
2. Incubate with shaking (90 r.p.m.) for 10 min at room temperature.
3. Add EDTA and Triton X-100 to final concentrations of 25 mM and 2%, respectively.
4. Sonicate with a tip sonicator (position 7 at 40% duty in pulse mode, 10 pulses followed by cooling in ice/salt bath; repeated three times on a Branson sonifier).
5. Centrifuge for 30 min at 50 000 *g* at 4 °C. Aliquots of the supernatant fractions should be preserved for analysis by SDS-PAGE (see *Figure 2*).
6. Add 10 ml of solution 2 to pellet and resuspend by sonication.
7. Centrifuge for 30 min at 50 000 *g* at 4 °C.

Protocol 9. *Continued*

8. Add 10 ml of solution 3 to pellet and resuspend by sonication.
9. Centrifuge for 30 min at 50 000 *g* at 4 °C.
10. Add 10 ml of solution 4 to pellet and resuspend by sonication.
11. Centrifuge for 30 min at 50 000 *g* at 4 °C.
12. Add 10 ml of solution 5 to pellet and resuspend by sonication.
13. Freeze aliquots of 1 ml at −80 °C for storage.

4.2 Import of purified precursor proteins into isolated mitochondria

To allow import into mitochondria, precursor proteins purified as inclusion bodies are solubilized using strong denaturants such as 7 M urea (solution 5, *Protocol 9*) or 6 M guanidinium-Cl (*Figure 2*) and are then diluted 100-fold into the import reaction. It has to be born in mind that the denatured

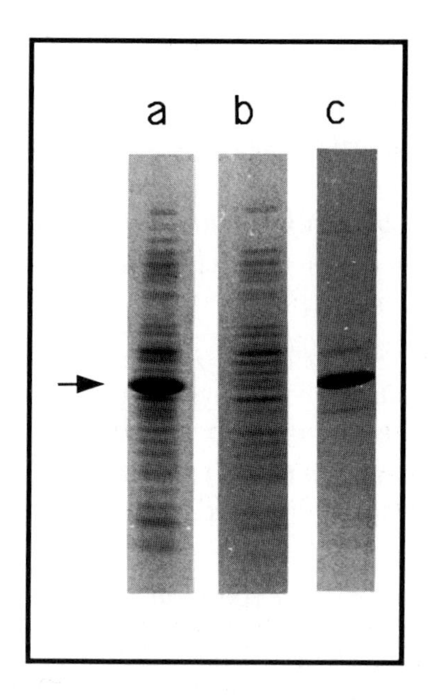

Figure 2. Purification of a fusion protein overexpressed in *E. coli.* Coomassie blue stained polyacrylamide gels of the following steps of the preparation: (a) *E. coli* cells lysed in SDS-containing buffer expressing the fusion protein b_2-[1–167]-DHFR as the major protein species (arrow); (b) cellular supernatant after pelleting the inclusion bodies (see *Protocol 9*, step **5**); (c) purified protein preparation (see *Protocol 9*, step **12**).

precursor protein undergoes refolding immediately upon dilution. In cases where high final concentrations of precursor (several mg/ml) are used, the precursor can form unproductive aggregates before import commences. We usually determine the concentration range of precursor at which aggregation poses a problem by spectrophotometrically recording the absorption at 320 nm (light scattering) in dilution experiments. In contrast to the import of radiochemical amounts of precursor from reticulocyte lysate, the import of chemical amounts of purified precursor protein requires the addition of Mg^{2+}-ATP. The amounts of ATP produced by mitochondria in the presence of respiratory chain substrates such as NADH are not sufficient to drive the import reaction. $MnCl_2$ can be included to provide sufficiently high concentrations of divalent cations to maintain the activity of the metal-dependent processing enzyme in the matrix of mitochondria.

Protocol 10. Import of a purified precursor protein

Materials

- import buffer: 250 mM sucrose; 20 mM MOPS, pH 7.2; 80 mM KCl; 2 mM $MgCl_2$; 2 mM $MnCl_2$; 2 mM NADH; 2 mM ATP; isolated mitochondria added from a 5 mg/ml stock to a final concentration of 1–2 mg/ml

Method

1. Incubate import buffer (200 μl) containing isolated mitochondria for 2 min at 25 °C.
2. Add 1–2 pmol purified mitochondrial precursor protein in urea or guanidinium buffer (maximal concentration of urea or guanidinium in the import reaction: 100 mM) and mix immediately.
3. Incubate for 15 min at 25 °C.
4. Cool for 5 min on ice.
5. Add proteinase K (20–100 μg/ml final concentration) and incubate for 20 min at 0 °C. Higher concentrations of proteinase K may be required when using chemical amounts of precursor protein.
6. Stop protease action by addition of PMSF to 1 mM and continue incubation for 5 min at 0 °C.
7. Re-isolate mitochondria by centrifugation for 10 min at 15 000 at 0 °C (Beckman JA-20 rotor with adaptors for microcentrifuge tubes) and resuspend in buffer suitable for further experimentation or solubilize in SDS-containing buffer for analysis by SDS–PAGE followed by fluorography or Western blotting using specific antibodies directed against the imported protein.

---

5. Translocation intermediates spanning contact sites

Sites of close contact between outer and inner mitochondrial membranes (contact sites) have been identified as the main entrance gates for precursor proteins moving from the cytosol into mitochondria (for review see reference 30). Translocation contact sites are functionally characterized by the reversible arrest of traversing polypeptide chains. To analyse structure and composition of contact sites, transport intermediates are accumulated in a two-membrane-spanning topology. Contact site intermediates are characterized by their accessibility to proteolytic cleavage on both sides of the membranes. (a) Precursor parts still outside the outer membrane are accessible to externally added proteases. (b) The amino-terminus of the precursor reaching into the matrix is proteolytically cleaved by the mitochondrial processing enzyme. To confirm the insertion of precursor proteins into contact sites it must first ascertained that the mitochondrial processing enzyme has not been released from the mitochondria and thereby gained access to the precursor protein. Cleavage of the pre-sequence of the precursor must therefore be dependent on the electric potential $\Delta\Psi$ across the inner membrane. Completion of transport through contact sites is a temperature-dependent process and requires a loosely folded conformation of the precursor protein. Precursors spanning the mitochondrial membranes are accumulated by several methods, all of which rely on stabilizing the folded state of the precursor protein (for review see reference 1). We present two procedures to generate contact site intermediates using the precursor of a fusion protein containing the 167 amino terminal residues of pre-cytochrome b_2 and the complete mouse dihydrofolate reductase (cytochrome b_2[1–167]-DHFR).

5.1 Translocation intermediates generated at low temperature

One procedure to accumulate transport intermediates during membrane translocation is to perform import into isolated mitochondria at reduced temperature (8–12 °C) (30). Under these conditions, the temperature-dependent unfolding of the DHFR part of the fusion protein is rate-limiting for translocation. Typical translocation intermediates of cytochrome b_2[1–167]-DHFR are produced that have the complete folded DHFR domain outside the outer membrane but reaching into the matrix space with the mitochondrial protein part where proteolytic cleavage occurs (*Figure 3*). In the absence of $\Delta\Psi$ no specifically processed, proteinase K-sensitive transport intermediates appear. The translocation arrest of these intermediates is reversible; raising the temperature to 25 °C allows completion of translocation into a proteinase K protected position within the mitochondria. The low temperature intermediates are kinetic intermediates; they allow only limited

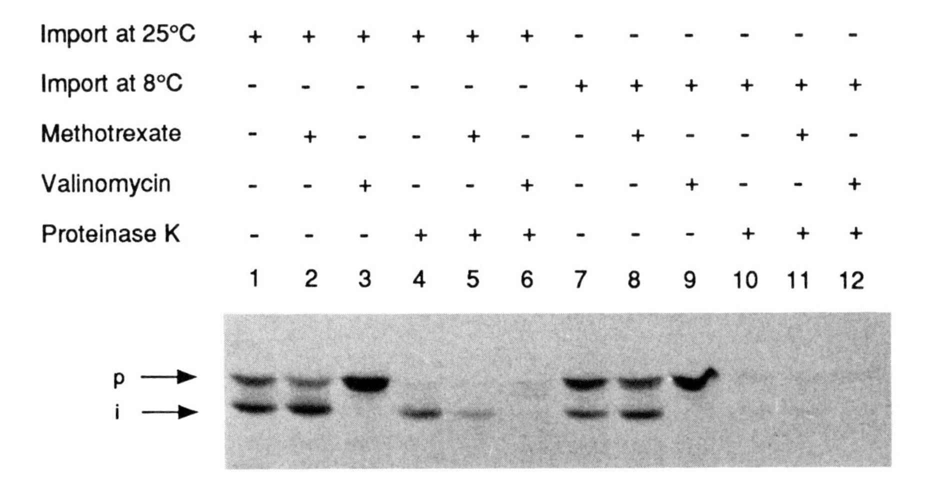

Figure 3. Import of b_2-DHFR fusion protein into isolated mitochondria. Mitochondria were incubated with reticulocyte lysate containing radiolabelled b_2-[1–167]-DHFR fusion protein. p, Precursor; i, intermediate-sized form cleaved by the matrix-localized processing enzyme. Import was performed either in the absence or in the presence (lanes 2, 5, 8, 11) of 100 nM methotrexate. Valinomycin (1 μM) was added to samples 3, 6, 9, and 12 to dissipate the membrane potential. Treatment of the mitochondria with proteinase K after import was performed to assay for complete membrane translocation of the precursor protein (lanes 4–6 and 10–12). When import was performed at 25 °C (lanes 1–6), precursor proteins were completely imported and were proteolytically processed (lanes 1 and 4). The processed form of the precursor was resistant to proteinase K treatment indicating that complete translocation had occurred. Addition of methotrexate (lanes 2 and 5) led to the accumulation of translocation intermediates spanning both mitochondrial membranes as shown by their accessibility to the processing peptidase in the mitochondrial matrix and to externally added proteinase K. Import at 8 °C for 15 min resulted in the accumulation of partially translocated proteins, both in the presence or in the absence of methotrexate (lanes 7, 8, 10, 11). Import and processing was blocked in the absence of a membrane potential (lanes 3, 6, 9, 12). (See *Protocol 5*).

manipulations of the mitochondria to further analyse interactions of the membrane-spanning precursor with components of the transport apparatus at contact sites. All manipulations of mitochondria carrying the intermediate must be performed at 0–4 °C.

5.2 Stable accumulation of translocation intermediates

An alternative approach to arrest transport intermediates in mitochondrial contact sites is to stabilize the tightly folded state of the carboxyl-terminal part of the precursor protein by binding of a specific antibody (31) or ligand (23, 32). In the case of the DHFR fusion proteins, this can be achieved by binding the strong folate antagonist methotrexate (MTX) to the DHFR-domain of the hybrid precursor protein (23, 32; Wienhues *et al.*, submitted for publication). Transport intermediates of cytochrome b_2-[1–167]-DHFR are generated in the presence of $\Delta\Psi$ when MTX (0.1–1 μM final concentration) is included in the import reaction (*Figure 3*). Although very stable, the block in translocation is reversible. Removal of MTX by re-isolation of mitochondria and suspension in MTX-free medium allows completion of import. In the presence of MTX, translocation intermediates are spanning contact sites in a defined structural arrangement allowing extensive manipulation of the mitochondria such as subfractionation in order to identify components associated with the translocating polypeptide.

6. Subfractionation of mitochondria

The sequential opening of the mitochondrial subcompartments intermembrane space and matrix is a useful tool in the investigation of the import pathway of precursor proteins. Only the determination of the correct intramitochondrial localization of imported precursors or of import intermediates will convincingly prove a postulated import route.

In principle, two methods are available for the subfractionation of mitochondria.

- The outer mitochondrial membrane can be ruptured selectively by controlled osmotic shock of mitochondria (33, 34, 20).
- By treatment of mitochondria with increasing concentrations of digitonin a sequential opening of the intermembrane space and matrix is achieved (34, 35, 20).

In order to test the submitochondrial localization of the imported protein, both methods can be combined with subsequent protease treatment and/or further separation into membranes and soluble components. The differential opening of the mitochondrial subcompartments can be monitored by analysing the release of subcompartment specific enzymes (e.g. cytochrome b_2 (37), adenylate kinase (37) or cytochrome c peroxidase (38) for the intermembrane space and fumarase or aconitase (39) for the matrix space) by measuring activities or by Western blotting of membrane and soluble fractions using antisera directed against marker proteins.

6.1 Subfractionation by osmotic swelling

Mitochondria are subjected to osmotic shock in a hypotonic medium which causes swelling of the matrix space. The outer membrane is more rigid than the inner membrane and is disrupted while the inner membrane remains intact. In this way the components of the intermembrane space are released or become accessible to added protease. The matrix components remain enclosed in the resulting mitoplasts which can be isolated by centrifugation or can be further subfractionated into inner membrane and matrix fractions (40).

Protocol 11. Osmotic swelling

Materials

- SEH medium: 250 mM sucrose; 1 mM EDTA; 10 mM Hepes–KOH, pH 7.4
- 20 mM Hepes–KOH, pH 7.4
- 2 M sucrose
- 3 M KCl

Method

1. Resuspend 100 μg mitochondria (re-isolated by centrifugation after import) in 10 μl SEH buffer (protein concentration 10 mg/ml).[a]
2. Add 155 μl 20 mM Hepes–KOH, pH 7.4 (protein concentration 0.55 mg/ml; sucrose concentration 15 mM), mix carefully, and incubate 15 min at 0 °C to allow for swelling.
3. Add 25 μl 2 M sucrose (final concentration 250 mM; restoration of isoosmotic conditions) and 10 μl KCl (3 M) (final concentration 150 mM; allows efficient release of intermembrane space enzymes). Mix gently but thoroughly and keep at 0 °C.
4. Use an aliquot corresponding to 35 μg of mitochondria to control the conversion to mitoplasts. Centrifuge at 20 000 *g* for 10 min at 2 °C. Remove the supernatant (do not discard) and resuspend the mitoplast pellet in SEH buffer. Measure enzyme activities of cytochrome b_2 as intermembrane space marker and of fumarase as matrix marker in the supernatant and mitoplast fractions.

[a] Controlled disruption of the outer membrane is not possible with mitochondria of *N. crassa*. For optimal fractionation freshly prepared yeast mitochondria must be used.

Protocol 12. Optional treatment of mitoplasts with proteinase K (19, 41)

1. Divide the reaction after swelling and restoration of isoosmosis into halves (*Protocol 11*, step **3**).
2. Add to one part 15–100 μg/ml proteinase K and incubate for 10–30 min at 0 °C.
3. Stop protease action by addition of 1 mM PMSF (also added in the control which lacks protease) and continue incubation for 5 min at 0 °C.

Protocol 12. *Continued*

4. Re-isolate mitoplasts by centrifugation at 20 000 *g* (Beckman JA20 rotor with adaptors for microcentrifuge tubes) for 10 min at 2 °C or precipitate total protein with 10% (final concentration) trichloroacetic acid (TCA) by centrifugation for 15 min in a cooled microcentrifuge. Wash TCA precipitates with 500 μl of acetone (precooled to −20 °C) and recentrifuge. Dry pellets under vacuum and dissolve them in SDS-containing buffer for SDS–PAGE.
5. Dissolve the mitoplast pellet or the protein pellet (after TCA-precipitation) in SDS-containing buffer and analyse by SDS–PAGE and fluorography.

Protocol 13. Optional separation of soluble and membrane fractions

1. After swelling of mitochondria and restoration of isoosmosis centrifuge the reactions for 10 min at 20 000 *g* (Beckman JA20 rotor with adaptors for microcentrifuge tubes).
2. Carefully remove the supernatant.
3. TCA-precipitate the supernatant proteins (see *Protocol 12*, step **4**). Dissolve precipitates and mitoplast pellets in SDS-containing buffer for analysis by SDS–PAGE.

6.2 Subfractionation by digitonin treatment

The differential sensitivity of the outer and inner membranes towards digitonin allows the sequential opening of intermembrane space and matrix by treatment of mitochondria with increasing concentrations of digitonin. At low concentrations, digitonin binds to ergosterol in the outer membrane (a derivative of cholesterol, which is absent from the inner membrane) and thereby destabilizes the outer membrane. As a result, the outer membrane barrier is disrupted thereby opening the intermembrane space while the matrix space remains intact. At higher concentrations, digitonin acts as detergent and solubilizes both outer and inner membranes. This results in the release of the contents of the matrix space. Digitonin fractionation works equally well with mitochondria of yeast and *N. crassa*.

Protocol 14. Digitonin treatment

Materials

- SEM buffer: 250 mM sucrose; 1 mM EDTA; 10 mM MOPS, pH 7.2

- SEMK buffer: SEM-buffer plus 100 mM KCl
- digitonin solutions of 0.25, 0.5, 1% (w/v) in SEMK

Note. Use digitonin solutions within 30 min of preparation. The commercially available digitonin should be recrystallized from hot ethanol to improve solubility.

Method

1. Resuspend 1 mg mitochondria (re-isolated after import) in 100 μl SEMK buffer (protein concentration 10 mg/ml).
2. Divide into 10 aliquots by pipetting 10 μl of the mitochondrial suspension to the bottom of the reaction tube (see *Table 1*).

Table 1. Composition of 10 aliquots in *Protocol 14*

	Volume (μl) in sample no.									
	1	**2**	**3**	**4**	**5**	**6**	**7**	**8**	**9**	**10**
Component										
Mitochondria	10	10	10	10	10	10	10	10	10	10
0.25% digitonin	0	4	6	8	10	0	0	0	0	0
0.5% digitonin	0	0	0	0	0	6	7	8	0	0
1% digitonin	0	0	0	0	0	0	0	0	6	10
SEMK	10	6	4	2	0	4	3	2	4	0
Final digitonin concentration (%)	0	0.05	0.075	0.1	0.125	0.15	0.175	0.2	0.3	0.5

3. Add digitonin from stock solutions as indicated in *Table 1*. Pipette to the wall of the reaction tube.
4. Adjust to a final volume of 20 μl by adding the required volume of SEMK (*Table 1*) to the bottom of the tube.
5. Spin for a few seconds in a microcentrifuge at 4 °C, mix gently but thoroughly, and incubate for 2 min at 0 °C.
6. Dilute 20-fold by addition of 400 μl SEMK buffer and mix gently (stops action of digitonin).
7. Divide each reaction into a 240 μl aliquot A (corresponds to 60 μg mitochondrial protein) and a 160 μl aliquot B (corresponds to 40 μg mitochondrial protein).
8. Use B to control for the extent of release of the intermembrane space and matrix contents: centrifuge at 20 000 *g* for 10 min at 2 °C; remove the supernatant (do not discard); resuspend the pellet in 160 μl SEMK; determine in each supernatant and pellet fraction the enzmatic activity of adenylate kinase (*N. crassa* mitochondria only) or cytochrome b_2 (yeast mitochondria only) and of fumarase.

Protocol 14. *Continued*

9. Use aliquot A for further experimentation as described in *Protocol 11*, step **4**.

Acknowledgements

We are grateful to Dr W. Neupert for valuable critical suggestions. Work in the authors' laboratory is supported by the Deutsche Forschungsgemeinschaft.

References

1. Pfanner, N., Hartl, F.-U., Guiard, B., and Neupert, W. (1987). *Eur. J. Biochem.* **169**, 289.
2. Hartl, F.-U., Pfanner, N., Nicholson, D. W., and Neupert, W. (1989). *Biochim. Biophys. Acta* **988**, 1.
3. Attardi, G. and Schatz, G. (1988). *Ann. Rev. Cell Biol.* **4**, 289.
4. Hartl, F.-U. and Neupert, W. (1990). *Science* **247**, 930.
5. Pfanner, N., Söllner, T., and Neupert, W. (1991). *Trends Biochem. Sci.* **16**, 63.
6. Weiss, H., von Jagow, G., Klingenberg, M., and Bücher, T. (1970). *Eur. J. Biochem.* **14**, 75.
7. Bradford, M. M. (1976). *Anal. Biochem.* **72**, 248.
8. Davenport, R. R. (1980). In *Biology and activities of yeast* (ed. F. A. Skinner, S. M. Passmore, and R. R. Davenport), p. 261. Academic Press, London.
9. Pelham, H. R. B. and Jackson, R. J. (1976). *Eur. J. Biochem.* **67**, 247.
10. Krawetz, S. A., Nanayama, A. S., and Anwar, R. A. (1983). *Can. J. Biol.* **61**, 274.
11. Marcu, K. and Dudock, B. (1974). *Nucl. Acids Res.* **1**, 1385.
12. Anderson, C. W., Straus, J. W., and Dudock, B. S. (1983). In *Methods in enzymology*, Vol. 101. (ed. R. Wu, L. Grossmann, and K. Moldave), p. 635. Academic Press, San Diego.
13. Gasior, E., Herrera, F., McLaughlin, C. S., and Moldave, K. (1979). *J. Biol. Chem.* **254**, 3970.
14. Moldave, K., and Gasior, E. (1983). In *Methods in enzymology*, Vol. 101 (ed. R. Wu, L. Grossmann, and K. Moldave), p. 644. Academic Press, San Diego.
15. Tai, P. C., Wallace, B. J., Herzog, E. L., and Davis, B. D. (1973). *Biochemistry* **12**, 609.
16. Rhoads, D. B., Tai, P. C., and Davis, B. D. (1984). *J. Bacteriol.* **159**, 63.
17. Krieg, P. A. and Melton, D. A. (1984). *Nucl. Acids Res.* **12**, 7057.
18. Davenloo, P., Rosenberg, A. H., Daum, J. J., and Studier, W. F. (1984). *Proc. nat. Acad. Sci., USA* **81**, 2035.
19. Pfanner, N., Tropschug, M., and Neupert, W. (1987). *Cell* **49**, 815.
20. Pfanner, N. and Neupert, W. (1986). *J. Biol. Chem.* **262**, 7528.
21. Schleyer, M., Schmidt, B., and Neupert, W. (1982). *Eur. J. Biochem.* **125**, 109.
22. Murakami, K., Amaya, Y., Takiguchi, M., Ebina, Y., and Mori, M. (1988). *J. Biol. Chem.* **263**, 18437.

23. Rassow, J., Guiard, B., Wienhues, U., Herzog, V., Hartl, F.-U., and Neupert, W. (1989). *J. Cell Biol.* **109**, 1421.
24. Schauder, B., Blöcker, H., Frank, R., and McCarthy, J. E. G. (1987). *Gene* **52**, 279.
25. Lanzar, M. and Bujard, H. (1988). *Proc. nat. Acad. Sci., USA* **85**, 8973.
26. Kane, J. F. and, Hartley, D. J. (1988). *Trends Biotechnol.* **6**, 95.
27. McCarthy, J. E. G. (1991). In *Advances in gene technology II*, (ed. P. J. Greenaway), pp. 145–75. JAI Press, London.
28. McCarthy, J. E. G. (1990). *Trends Genet.* **6**, 78.
29. Neupert, W., Hartl, F.-U., Craig, E. A., and Pfanner, N. (1990). *Cell* **63**, 447.
30. Pfanner, N., Rassow, J., Wienhues, U., Hergersberg, C., Söllner, T., Becker, K., and Neupert, W. (1990). *Biochim. Biophys. Acta* **1018**, 239.
31. Schleyer, M. and Neupert, W. (1985). *Cell* **43**, 339.
32. Eilers, M. and Schatz, G. (1986). *Nature* **322**, 228.
33. Ohba, M. and Schatz, G. (1987). *EMBO J.* **6**, 2117.
34. Hartl, F.-U., Schmidt, B., Wachter, E., Weiss, H., and Neupert, W. (1986). *Cell* **47**, 939.
35. Stuart, R., Nicholson, D. W., and Neupert, W. (1990). *Cell* **60**, 31.
36. Appleby, C. A., and Morton, R. K. (1959). *Biochem. J.* **71**, 492.
37. Schmidt, B., Wachter, E., Sebald, W., and Neupert, W. (1984). *Eur. J. Biochem.* **144**, 581.
38. Djavadi-Ohaniance, L., Rudin, Y., and Schatz. G. (1978). *J. Biol. Chem.* **253**, 4402.
39. Racker, E. (1950). *Biochim. Biophys. Acta* **4**, 211.
40. Daum, G., Böhni, P. C., and Schatz, G. (1982). *J. Biol. Chem.* **257**, 13028.
41. Scherer, P. E., Krieg, U. C., Hwang, S. T., Vestweber, D., and Schatz, G. (1990). *EMBO J.* **9**, 4315.

7

Techniques in nuclear protein transport

COLIN DINGWALL

1. Introduction

The transport of macromolecules across the nuclear envelope occupies a central place in the physiology of the cell. Interest in the study of this process has been relatively slow in developing but it now represents a major area of investigation. Recently there have been a number of demonstrations of regulated nuclear entry of proteins involved in cell cycle control, cellular transformation, and determination of cell fate in development. A variety of biochemical events are thought to bring about this regulated entry. These include protein phosphorylation or proteolysis to release a nuclear protein from an attachment site in the cytoplasm or reveal a cryptic nuclear targeting sequence. Unravelling the molecular basis of this control will require reliable techniques and many of those in current use are described in this chapter.

2. Microinjection—amphibian oocytes

The discovery that nuclear proteins contain a permanent signal specifying entry and accumulation in the cell nucleus was made possible by the availability of techniques for the microinjection of proteins into cells and the simple analysis of the fate of the injected probes. The cells used were predominantly amphibian oocytes which are large cells (1 mm diameter) that can be microinjected easily. They have an extremely large nucleus (100 μm in diameter) which is readily isolated, allowing the analysis of individual cells.

This remains the system of choice for many studies of transport into and out of the nucleus and in this section protocols are provided for the removal of oocytes from the African clawed toad, *Xenopus laevis*, their maintenance in culture, and microinjection and analysis.

Protocol 1. Removal and maintenance of *Xenopus* oocytes

Materials

- one fully grown female *Xenopus* frog
- ethyl *m*-aminobenzoate
- hibitane (10% solution of chlorhexidine gluconate in 70% ethanol)
- Barth's modified saline, BMS: 88 mM NaCl, 1 mM KCl, 2.4 mM $NaHCO_3$, 0.33 mM $Ca(NO_3)_2$, 0.41 mM $CaCl_2$, 0.82 mM $MgSO_4$, 10 mM Hepes, 10 mg/l streptomycin sulphate, 10 mg/l benzyl penicillin[a]

Method

1. Anaesthetize the frog by immersion in 200 ml water containing 0.5 g ethyl *m*-aminobenzoate for 15–30 min in a plastic box with a lid.
2. Rinse the anaesthetized frog under cold running water and place it on its back on damp paper.
3. Wipe ventral surface with hibitane and using sharp scissors make an incision about 1 cm long in the abdomen first cutting through the skin and then through the muscle layer (*Figure 1*).

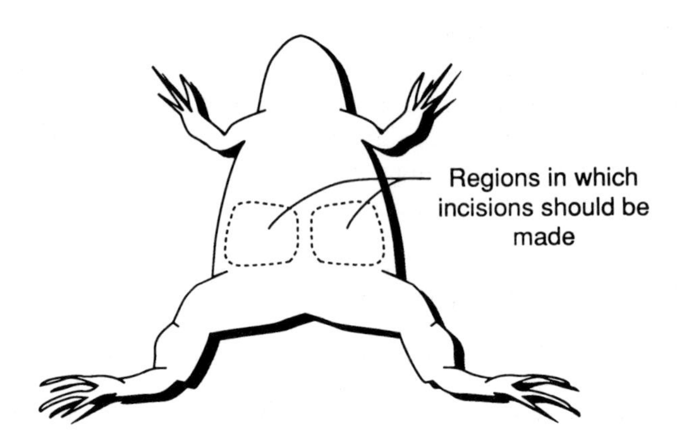

Figure 1. Sites for incisions to remove ovaries from *Xenopus laevis* female.

4. With fine forceps gently pull the lobes of the ovary through the incision and cut off pieces until enough oocytes have been collected (a gram of ovary will contain about 500 fully grown oocytes).
5. Sew incision with a non-self-digesting surgical thread through skin and muscle. Two or three stitches are required.

6. Allow the frog to recover in a small tank held at an angle with the frog lying in a small volume of water (but not fully immersed). The back of the frog may be covered with damp tissue.
7. When the frog has recovered (~2 h), place in tap water and keep separate from remainder of frog colony.
8. Remove stitches after 10 days.[b]
9. Place the lobes of ovary in a Petri dish of BMS and tease into clumps of about 20 oocytes using watchmaker's forceps.
10. Wash oocytes free of blood and remnants of broken oocyte using several changes of BMS and store oocytes in fresh BMS in Petri dishes such that clumps of oocytes are not in direct contact.[c]
11. Store dishes of oocytes at ~ 19 °C in Perspex box with damp tissue.

[a] Magnesium sulphate is added last and the pH is then brought to 7.5 with sodium hydroxide and the solution made to volume. Ficoll to a final concentration of 0.1% may be added. This increases the storage time of oocytes and aids in their recovery after microinjection. BMS can be made up as 10 × stock, sterilized, and stored for long periods at 4 °C or frozen. Antibiotics are added when diluted for use.

[b] Frogs can be re-used at approximately 6-month to 1-year intervals.

[c] Oocytes can be stored for several days if the BMS is changed twice daily and dead or necrotic oocytes are removed. Such oocytes are easily identified by the loss or 'marbling' of the pigment of the animal pole.

2.1 Microinjection pipettes

Microinjection pipettes are produced from glass capillaries, generally by two successive steps of melting and drawing. The first 'pull' may be carried out manually, but the second 'pull' requires a pipette puller of some description. The final step is the forging of the pipette tip to produce a sharp bevelled end (*Figure 2*). A number of pipette pullers are now available commercially which

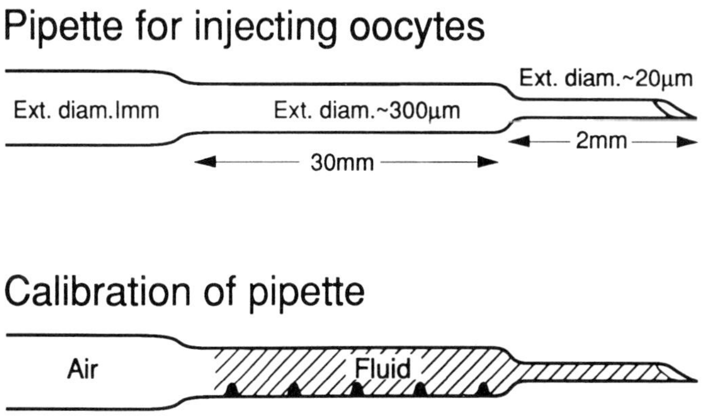

Pipette for injecting oocytes

Ext. diam.~20μm
Ext. diam.lmm
Ext. diam.~300μm
2mm
30mm

Calibration of pipette

Air
Fluid

Figure 2. Dimensions of microinjection pipettes.

allow the heat and strength of the 'pull' to be varied. The crucial factor is that the pipette should be sufficiently sharp to enter the oocyte easily and not cause leakage of cytoplasm when withdrawn. Too fine a pipette can become clogged with cytoplasm. The method described here produces pipettes that are easily calibrated and the meniscus of the fluid being injected can be followed under the microscope during injection.

Protocol 2. Microinjection pipette production and calibration

Materials

- BDH hard glass capillaries (Cat. no. 32124) or capillaries of similar dimensions (100 mm × 1.5 mm o.d. × 1 mm i.d)
- pipette puller such as that available from Research Instruments Ltd
- Microforge such as that available from the same manufacturer or from De Fonbrune Metallurgical Service Laboratories Ltd., Reliant Works

Method

1. Heat the central 1 cm of the capillary in a Bunsen flame until soft, remove from the flame, and pull out to a total length of approximately 30–40 cm. Capillaries are chosen that have a parallel (non-tapering) central, pulled section with an outer diameter of 1 mm or slightly less.
2. Break off the pipette to give an overall length of 10–12 cm to allow a second pull at 2–3 cm from the shoulder produced by the first 'pull'.
3. The second pull is carried out using a suitable commercial pipette puller. The end is then broken with fine watchmaker's forceps and, using a suitable microforge, the end is forged to produce a sharp bevelled end (*Figure 2*).
4. The pipette is attached to the microinjection apparatus (see *Protocol 3*) for calibration. Pipette 1 μl of water or other suitable liquid (1 × BMS) on to a piece of Parafilm or the lid of a microcentrifuge tube and place the tip of the pipette in the drop of liquid.
5. Draw the liquid up into the pipette; routinely the liquid occupies about 2 cm of the pulled pipette.
6. With a fine pointed felt-tip pen make equally spaced marks on the pipette from the meniscus of the fluid to the tip of the pipette (*Figure 2*). Marks at 1-mm intervals therefore correspond to approximately 50-nl volumes. This is the volume routinely injected into the cytoplasm of oocytes for nuclear transport studies. Injection of significantly larger volumes into the oocyte can be deleterious. Pipettes for microinjection into the nucleus are made in the same way but from smaller capillaries.

2.2 Microinjection techniques

Probe molecules may be injected into the nucleus or cytoplasm of the *Xenopus* oocyte using essentially the same basic technique. With practice cytoplasmic injections can be rapidly performed on a large number of oocytes. For nuclear injections two basic methods are in use. The first relies upon the fact that the nucleus lies in a fixed position at the animal pole. Injections can be directed towards the nucleus and the pipette allowed to pepetrate to the correct depth. The technique can be practiced by injecting solutions of Trypan blue which does not diffuse from the site of injection. After trial injections the oocytes can be dissected to determine the frequency of successful injections. With practice a success rate of greater than 80% can be achieved.

Alternatively, oocytes can be centrifuged gently. This forces the yolk platelets to the bottom of the oocyte and the nucleus rises. The position of the nucleus is then evident as a slighly raised, paler patch at the animal pole.

Protocol 3. Cytoplasmic injection of *Xenopus* oocytes

Materials

- oocytes[a]
- Agla type syringe with micrometer screw and sprung plunger (Singer Instruments)
- micromanipulator that reduces hand movements by a factor of five (Microdissector of Singer Instruments)
- microinjection pipettes
- stereomicroscope
- silicone tubing
- liquid paraffin

Method

1. Connect the silicone tubing to the Agla syringe and fill entirely with liquid paraffin.
2. Connect a calibrated microinjection pipette to the tubing and attach to micromanipulator; the liquid paraffin should enter the pipette to a distance of about 0.5 cm.
3. Place 1 μl of the liquid to be injected on a piece of Parafilm or the lid of a microcentrifuge tube. Place the tip of the pipette in the drop of liquid and using the micrometer screw draw the liquid into the pipette.
4. Cut a 1-cm strip of glass from a microscope slide and stick it across the centre of a second slide with a smear of silicone grease.

Protocol 3. *Continued*

5. Tease out a clump of 6–8 oocytes and place on the slide against the ledge formed by the glass strip. (Survival of injected oocytes is improved if they are injected and incubated in groups of 6–8.)
6. Using watchmaker's forceps arrange the oocytes as far as possible in a row with the vegetal pole showing. Blot off excess saline with a piece of tissue paper.
7. Bring the pipette tip to an oocyte and inject, aiming for the centre of the oocyte and avoiding the nucleus. A sharp pipette should enter the oocyte easily; a blunt pipette may require extra pressure or a slight stabbing motion to penetrate the oocyte. If cytoplasm spills out of the oocyte, discontinue use of the pipette.
8. With the pipette tip in the oocyte, slowly turn the micrometer screw observing the meniscus of the liquid in the pipette. Stop when the required volume is injected.
9. Withdraw the pipette and move on to the next oocyte. When all oocytes in a group have been injected place in BMS. It should take less than 1 min to inject a small group of oocytes.

[a] Mature oocytes have a darkly pigmented hemisphere, the animal pole, and a less pigmented hemisphere, the vegetal pole, which appears grey-yellow due to the presence of yolk platelets in the cell. These cells are arrested in the first meiotic prophase and the nucleus is positioned roughly 50 μm below the surface of the oocyte at the animal pole. Stage-6 oocytes are preferred for microinjection; characteristically they have a dark pigmented animal pole, a yellow/grey vegetal pole; and a slightly lighter coloured equatorial band.

Protocol 4. Direct nuclear injection of *Xenopus* oocytes

Materials

As for *Protocol 3*, but the pipettes are made from narrower bore capillary tubing and calibrated to deliver 10–15 nl volumes.

Method

1. The microinjection apparatus is set up as described for cytoplasmic injections (*Protocol 3*).
2. Arrange oocytes on a microscope slide as in *Protocol 3* but with the animal pole directed towards the microinjection pipette.
3. Bring the pipette up to the oocyte and inject to a depth of about 0.1–0.2 mm. The pipette tip can be marked with a waterproof fine felt-tipped pen at the appropriate point.
4. Inject 10–15 nl as for cytoplasmic injections.
5. When all oocytes in a group have been injected transfer to BMS.

Protocol 5. Centrifugation of oocytes for nuclear injection[a]

Materials

- as for *Protocol 3*
- 5 cm plastic Petri dish
- nylon mesh with 1-mm mesh size

Method

1. Fix the nylon mesh to the bottom of the Petri dish with adhesive or suitable solvent (e.g. chloroform).
2. Using watchmaker's forceps dissect individual oocytes from the groups stored in BMS.
3. Place the oocytes on the grid, animal pole uppermost, in a minimal volume of BMS.
4. Centrifuge at 500 *g* for 10 min. A circular discoloration appears where the nucleus has displaced the pigment.
5. With the oocytes in place in the Petri dish inject into the nucleus which now lies just below the surface of the oocyte at the site of the discoloration.
6. Transfer injected oocytes from the dish to fresh BMS.

[a] A high success rate of nuclear injections is achieved using this method. In contrast to non-centrifuged oocytes the nucleus can only be isolated after trichloroacetic acid (TCA) fixation of the oocyte (see section 2.3).

2.3 Isolation of nuclei

The nucleus can be removed manually by gently tearing the oocyte open at the animal pole using two pairs of watchmaker's forceps. The nucleus is visible as a clear sphere lying in the yellow cytoplasm. It can be separated from the oocyte using the forceps and picked up in a small volume (1–3 µl) using an automatic pipette with a cut-off tip. The nucleus may be washed free of adhering cytoplasm by gently pipetting it back and forth in the pipette. The cytoplasm is picked up in a similar manner in a 2–5 µl volume, vortexed in a suitable buffer and centrifuged to remove the yolk. The clear supernatant is then analysed. Sodium dodecyl sulphate (SDS) sample buffer should not be added to whole cytoplasm as this disrupts the yolk platelets and the liberated yolk proteins interfere with SDS gel electrophoresis.

Alternatively, the oocytes can be immersed in 20% TCA for 5–10 min. The oocyte can then be dissected and the fixed nucleus readily isolated.

3. Colloidal gold

Colloidal gold particles are electron-opaque, allowing easy visualization in the electron microscope and visualization at the light microscope level by silver enhancement techniques (1). The particles physically adsorb on to their surface molecules such as proteins, peptides, carbohydrates, and nucleic acids. Consequently, colloidal gold has proved to be one of the most valuable probes used in the study of macromolecular transport across the nuclear envelope. For example, the definitive evidence that transport occurs through the nuclear pores, that nuclear entry occurs via a two-step mechanism, and that a single-pore complex can apparently simultaneously export an RNA molecule and import a protein molecule all came from the use of colloidal gold probes (reference 2 and references therein).

3.1 Preparation of gold solutions

Colloidal gold solutions are prepared by reduction of chloroauric acid with either white phosphorus, sodium ascorbate, or sodium citrate. In some protocols the amount of reducing agent can be varied to adjust the mean size of the gold particles in the solution. All glassware must be scrupulously clean and siliconized.

Protocol 6. Reduction of chloroauric acid with sodium citrate

Materials

- chloroauric acid
- tri-sodium citrate
- NaCl
- water that has been double distilled or purified by reverse osmosis and ion exchange. A final ultrafiltration step through a 0.2-μm filter is recommended

Method

1. Dissolve 1 g of tri-sodium citrate in 100 ml ultrapure, filtered (0.2 μm) water.
2. Dissolve 20 mg chloroauric acid in 200 ml ultrapure, filtered (0.2 μm) water.
3. Add an appropriate volume of tri-sodium citrate solution to the chloroauric acid solution to produce particles of the required size.[a]
4. Mix the solutions thoroughly by swirling in a conical flask and allow to stand for 5 min at room temperature during which time the solution loses its yellow colour.

5. Heat the solution rapidly to 70 °C under reflux, stirring as slowly as possible using a magnetic stirrer (use clean Teflon- or glass-covered stir bars) and bring close to boiling point slowly but do not allow to boil. The solution goes through a number of colour changes at this stage from blue-purple to the final crimson red colour. This process may take up to 10 min.
6. When the final colour is achieved, allow to simmer gently for 30 min. The solution deepens to a burgundy red.[b]
7. Allow the gold solution to cool slowly, stirring gently. When cool transfer to a clean screw-topped flask and store at 4 °C.[c]

[a] For ~ 12-nm diameter particles use 4.6 ml citrate; for ~ 20-nm diameter particles use 3.6 ml citrate; for ~ 30-nm diameter particles use 3.0 ml citrate; for ~ 50-nm diameter particles use 2.0 ml citrate.

[b] Any blueness indicates aggregation and the preparation must be discarded. About 60% of preparations are successful. Those that fail do so due to overvigorous heating and stirring and use of insufficiently pure, filtered water.

[c] Colloidal gold solutions prepared in this way are stable for up to 6 months at 4 °C.

Protocol 7. Reduction with phosphorus

Materials

- white phosphorus
- diethyl ether
- chloroauric acid
- ultrapure water

Method

1. Wash the large pieces of white phosphorus in diethyl ether to remove oil and cut into small pieces and transfer to a 250-ml conical flask containing 200 ml diethyl ether. Use enough phosphorus to cover the bottom of the flask and stopper tightly. **Leave for at least 1 month**. Poor results are obtained using phosphorus-saturated ether less than 1 month old.
2. Add 15 mg of chloroauric acid to 120 ml ultrapure filtered water followed by 3 ml of 0.2 M potassium carbonate.
3. Prepare a phosophorus–ether mixture by mixing four parts ether with one part phosphorus-saturated ether and add 1 ml of the phosphorus–ether mixture to the chloroauric acid/potassium carbonate solution.
4. Allow to stand at room temperature for 15 min and then boil for 5 min to produce a wine-coloured solution.

Protocol 7. *Continued*

5. Allow to cool slowly by standing the flask in a Styrofoam box at room temperature. Store the cooled solution at 4 °C.[a]

[a] Solutions produced by this method tend to contain particles 4.6–5.8 nm in diameter. After stabilization with the desired protein, nucleic acid, or peptide the gold solution can be sedimented through a sucrose gradient to prepare a fraction containing particles of a desired size (see Section 3.3).

Protocol 8. Reduction with sodium ascorbate

Materials

- potassium carbonate
- chloroauric acid
- sodium ascorbate
- ultrapure water

Method

1. Add 1.0 ml of 1% chloroauric acid and 1.0 ml of 0.1 M potassium carbonate to 25.0 ml ultrapure, filtered water at 0 °C.
2. Stir gently and add 1.0 ml of 0.7% sodium ascorbate in ultrapure water. Larger particles are produced by carrying out this step at room temperature. The solution becomes purple-red immediately upon the addition of the ascorbate.
3. The volume is adjusted to 100 ml with ultrapure Millipore-filtered water and heated to boiling to produce a red solution. This method produces particles with an average diameter of about 15 nm.

3.2 Protein conjugation

The same basic techniques can be used for stabilization with peptides and proteins. However, peptides having a high or low isoelectric point can be difficult to coat directly on to gold particles. This can be circumvented by cross-linking the peptide to a carrier protein such as bovine serum albumin and the protein–peptide conjugates can be used to stabilize the gold solution. The exact amount of coating agent should be determined for each individual gold preparation and can vary considerably. For example to coat 1 ml of a gold preparation requires 6 μl (60 μl of 0.1 mg/ml) nucleoplasmin, while 60 μg (60 μl of 1 mg/ml) of serum albumin is needed to stabilize the same gold preparation.

Protocol 9. Coating of colloidal gold with proteins

Materials

- gold solution
- protein solution in a low ionic strength buffer at a pH close to the pI of the protein
- NaCl

Method

1. Dispense 100-μl aliquots of gold solution into clean tubes and add an aliquot of a suitable buffer stock[a] to give a final concentration of 50 mM. The final pH of the solution must be at or on the basic side of the isoelectric point of the protein or peptide.
2. Add aliquots of protein to the gold solutions to give a range of protein concentrations from 10 to about 200 μg/ml.
3. Mix the protein and gold solutions gently (do not vortex) and allow to stand at room temperature for 5–15 min.
4. The solutions are 'challenged' by the addition of 10 μl 1% NaCl to each.[b]

[a] There is no standard buffer but most commonly used buffers are compatible with colloidal gold. A buffer system is chosen that gives a pH at or near the isoelectric point of the material to be coated on the gold.

[b] A stabilized solution (gold particles successfully coated) remains red; a solution that is not stabilized turns blue instantly because of aggregation of the gold particles. Choose the lowest concentration of protein that stabilizes the gold against challenge by NaCl and then scale up. Routinely, 10 ml of gold solution is stabilized using the appropriate amount of protein. A small aliquot is then challenged with NaCl to ensure that stabilization has been successful. The gold solution is then concentrated by centrifugation (see Section 3.3).

3.3 Concentration and fractionation of coated gold particles

Gold solutions produced by the phosphorus method are centrifuged at 40 000 r.p.m. for 30 min in a Beckman SW-60 rotor. Gold solutions produced by the citrate or ascorbate methods are centrifuged at 15 000 r.p.m. for 30 min in a Sorvall RC-5B centrifuge in a SS-34 rotor. The solution forms a loose pellet at the bottom of the tube and the supernatant is carefully withdrawn with a Pasteur pipette. The solution may be resuspended in 150 mM NaCl, 2.5 mM KCl, 10 mM phosphate buffer, pH 7.2 and recentrifuged. The loose part of the second pellet is resuspended in 0.2–1 ml of resuspension in buffer which also contains 50 μg/ml bovine serum albumin (BSA) or 100 μg/ml PVP (polyvinyl pyrollidone; average molecular weight 360 000, from Sigma) to help stabilize the gold solution. The stabilized gold preparation is ready for use and is stable for weeks in a tightly stoppered tube at 4 °C.

In order to prepare uniformly sized subfractions the gold solution produced after the second centrifugation step is layered on a 10–30% linear sucrose or glycerol gradient in a Beckman SW40 tube (10.5 ml, 8 cm length). Gradients are centrifuged at 41 000 r.p.m. for 45 min for gold prepared by the phosphorus method or at 20 000 r.p.m. for 30 min for gold prepared by the ascorbate/citrate methods. 1-ml fractions are collected and examined by electron microscopy. Desired size fractions are dialysed against resuspension buffer and concentrated as described in the preceding paragraph.

3.4 Nucleic acid conjugation

Colloidal gold particles have been successfully coated with a number of different nucleic acids (2). The basic procedure is similar to that for peptides and proteins, and the success of coating is assayed using the NaCl challenge technique described in *Protocol 9*.

tRNA, poly A (3500 bases), poly (dA) (500 bases) are dissolved in 10 mM KCl, 7.2 mM K_22HPO_4, 4.8 mM KH_2PO_4, pH 7.0. 5SRNA is dissolved in 1 mM Tris-HCl, pH 7.5; 10 mM NaCl, 0.1 mM $MgCl_2$. The method is exactly as described in *Protocol 9* for the stabilization with proteins.

4. Peptide conjugation and protein labelling

One of the remarkable properties of nuclear targeting sequences is their ability, as peptides, to direct molecules to the nucleus when they are chemically cross-linked to non-nuclear proteins (3). A comparison of the transport properties of a number of peptide sequences has led to the proposal of a four amino-acid consensus sequence, lys-arg/lys-X-arg/lys, that specifies peptide-mediated nuclear targeting in which X may be lysine, arginine, proline, valine, or alanine but not asparagine (4). It should be noted that in this form the immediately adjacent sequence context has been shown to affect significantly the ability of a given peptide to direct a protein to the cell nucleus.

In general, basic amino acids and predominantly lysine residues occur frequently in nuclear targeting sequences. It is important to point out that many of the cross-linking and labelling techniques are based upon the chemistry of reactions with these basic side chains. In view of this it is important to control carefully the extent of modification of lysine residues that occurs during conjugation. For example, it is possible to label nucleoplasmin with lysine-modifying reagents and not inactivate the signal as long as concentrations are limited so that only one of the five monomers in the pentamer is labelled. On the other hand, in the SV40 nuclear targeting sequence one lysine is crucial for efficient nuclear targeting and similar reactions could inactivate the signal. In the following protocols, labelling and cross-linking is achieved through thiol groups which so far have not been implicated as important amino acids in any nuclear targeting sequence. In peptide nuclear targeting sequences the most common and most successful approach is to have, in addition to the nuclear targeting sequence, a tyrosine residue that can be radio-iodinated, and a free cysteine residue for cross-linking.

Protocol 10. Peptide conjugation using a heterobifunctional cross-linking reagent

Materials

- *m*-maleimidobenzoyl-*N*-hydroxysuccinimide ester (MBS; Pierce Chemical Co.)
- 50 mM phosphate buffer, pH 7.0
- carrier protein (e.g. BSA)
- peptide with terminal tyrosine residue and cysteine residues

Method

1. Dissolve carrier protein in phosphate buffer to a concentration of 5–10 mg/ml.
2. Slowly add up to a 20 molar excess of MBS in dimethyl formamide (DMF) to the carrier protein solution. Routinely this is achieved by adding 5–50 μl of a 10 mg/ml stock solution of MBS to the protein solution.[a]
3. Incubate for 1 h at 23 °C.
4. Remove unreacted MBS by gel filtration through Sephadex G50 equilibrated with 50 mM phosphate buffer, pH 7.0.
5. Mix a 50 molar excess of unlabelled synthetic peptide containing 1 μg of iodinated peptide at approximately 3×10^6 c.p.m. in 100 mM phosphate buffer, pH 6.0 with the derivatized carrier protein.
6. Incubate for 3 h at 23 °C.[b]
7. Dialyse against phosphate buffer to remove unconjugated peptide.
8. Analyse conjugate by SDS gel electrophoresis and by gamma counting to determine the coupling ratio. Routinely 3–15 peptides are cross-linked per carrier molecule.[c]

[a] The molar ratios of the reactants should be varied to obtain the desired coupling ratios.
[b] DMF to a final concentration of 20% may be added to increase solubility.
[c] Conjugated proteins may be concentrated by centrifugation through a Centricon 30 microconcentrator.

Protocol 11. Fluorescent labelling through SH groups

Materials

- iodoacetamidofluorescein in water
- protein labelling buffer: 100 mM Tris HCl, pH 8.0; 10% glycerol; 1 mM EDTA; 50 mM NaCl; 0.1 mM phenylmethylsulphonyl fluoride (PMSF)

Method

1. Dialyse protein against labelling buffer and dilute protein to a final concentration of 50 μg/ml.
2. Add stock iodoacetamidofluorescein to 100 μl protein to give a final concentration of < 200 mM.
3. Purge tube with nitrogen, seal, and incubate for 16 h at 16 °C in the dark.
4. Remove label by dialysis or gel filtration through Sephadex G75 equilibrated with a suitable buffer.

5. Protein transport *in vitro*

The transport of proteins and RNA across the nuclear envelope has been successfully reproduced *in vitro*. For the study of protein import there are two broad classes of system—those systems in which *in vitro* nuclear protein import has been reported using purified nuclei (5) or nuclear envelope vesicles (6) in a defined buffer system; and biochemically more complex systems that utilize nuclei added to, or reconstituted in, amphibian egg extracts (7) or permeabilized whole cells (8). Specific and efficient transport has been reported in both types of system. A requirement for cytoplasmic factors has been demonstrated in the extract and whole cell systems, while there appears to be no such requirement for transport in isolated nuclei.

Generally speaking, the extract-based systems have gained more widespread acceptance, while the systems using purified nuclei and nuclear envelope vesicles have been relatively less popular. Perhaps by *in vitro* complementation of these systems a consistent picture will emerge concerning the requirement for additional factors. Protocols are provided for both types of system.

The molecular probes used are radiolabelled or fluorescently labelled nuclear proteins such as nucleoplasmin, and non-nuclear proteins, such as serum albumin, are used as negative controls. More recently, protein transport has been studied in mouse liver nuclei (9) and yeast nuclei (10) using ^{35}S-labelled proteins produced by *in vitro* translation of a synthetic RNA transcript of the appropriate cDNA clone using conventional techniques.

5.1 Transport in amphibian egg extracts

The systems available for the study of transport *in vitro* are all based on the pioneering work of Lohka and Masui (11). These authors showed that a preparation of amphibian (*Xenopus*) egg cytoplasm is able to decondense sperm pronuclei and assemble a membrane-enclosed nucleus. These nuclei replicate their DNA and undergo the initial stages of mitosis. This system has also been used to investigate the steps in nuclear envelope assembly.

In the study of protein import, nuclei assembled around bacteriophage DNA, sperm DNA, sperm pronuclei, rat liver nuclei, and nuclei from cells grown in culture have all been shown to import proteins efficiently when incubated in egg extracts.

The following protocols detail the production of extracts, nuclei, and the methods for studying transport.

Protocol 12. Production of activated *Xenopus* eggs

Materials

- 1 female *Xenopus* frog
- Folligon (follicle-stimulating hormone, Intervet Laboratories)
- chorionic gonadotrophin (Chorulon, Intervet Laboratories)
- high salt Barth's saline: 110 mM NaCl; 2 mM KCl; 1 mM $MgSO_4$; 0.5 mM Na_2HPO_4; 2 mM $NaHCO_3$; 15 mM Tris-HCl, pH 7.4
- 2% cysteine hydrochloride, pH 7.8 in 1/10 th Barth's solution
- 20% Barth's modified solution: 18 mM NaCl; 0.2 mM KCl; 0.5 mM $NaHCO_3$; 2 mM Hepes/NaOH, pH 7.5; 0.15 mM $MgSO_4$; 0.5 mM $Ca(NO_3)_2$; 0.1 mM $CaCl_2$

Method

1. Prime female *Xenopus* frogs by injection of 100 units Folligon 4–7 days prior to chorionic gonadotrophin injection.
2. Inject 500 IU of chorionic gonadotrophin into the dorsal lymph sac. Egg-laying should begin about 12 h later. Alternatively, an initial injection of 200 IU of chorionic gonadotrophin can be given with a further 400–500 units 6–8 hours later. Egg-laying should commence about 6 h after the second injection.
3. Collect eggs in high salt Barth's saline. The whole frog is immersed in high salt Barth's saline for the period of laying (normally about 24 h or overnight), then transferred to water. The frog tank is fitted with a perforated tray so that the eggs fall through to the bottom of the tank. Eggs are harvested from the high salt Barth's saline.
4. Transfer eggs to a solution of 2% cysteine in 20% modified Barth's

Protocol 12. *Continued*

solution and swirl the eggs gently in the buffer. This removes the jelly coat from the eggs and takes approximately 10–20 min. During this process examine the eggs under a dissecting microscope and remove all damaged and prematurely activated eggs using a cut down Pasteur pipette.

5. Wash de-jellied eggs several times in 20% modified Barth's solution.
6. Activation can be achieved by electric shock or by the use of a calcium ionophore. Both procedures are described here but a greater proportion of activated eggs are produced with the calcium ionophore.
   - Activate eggs by electric shock (80 V for 1 sec) in 1/10th BMS. This is most conveniently carried out in a flat bed gel apparatus. Activated eggs are identified by their relative rigidity, flattened appearance, and contraction of the pigmented area. Again, damaged and non-activated eggs are removed.
   - Activate eggs by incubation in 0.5 mg/ml calcium ionophore A23187 (Sigma) in 20% Barth's modified solution for 5 min. Wash in 20% Barth's modified solution and incubate for a further 15 min. Remove all damaged and non-activated eggs.

Protocol 13. Preparation of egg extract

Materials

- activated eggs
- extraction buffer: 250 mM sucrose; 50 mM KCl; 1 mM dithiothreitol (DTT); 2.5 mM $MgCl_2$; 2 μg/ml cycloheximide; 5 μg/ml cytochalasin B; 3 μg/ml leupeptin. (All chemicals from Sigma Chemical Co.)

Method

1. Rinse eggs in ice-cold extraction buffer within 15–20 min of activation.
2. Transfer to 5-ml tube and allow eggs to settle.
3. Remove as much of the supernatant as possible and lyse eggs by centrifugation at 9000 *g* (7000 r.p.m. in a Sorvall HS-4 rotor) for 20 min at 4 °C. After centrifugation four major layers or fractions are visible. A dense insoluble plug of yolk platelets and pigment granules forms a pellet at the bottom of the tube. On top of this layer is a yellow-brown cytoplasmic layer (heavy ooplasmic fraction) above which is a translucent layer (light ooplasmic fraction). On top of this is a layer of lipid.
4. Collect the heavy ooplasmic fraction with a wide-mouthed Pasteur pipette and centrifuge in a Beckman SW 50Ti rotor at 9000 r.p.m. for 5 min at 4 °C in order to remove residual debris.
5. Freeze extract in small aliquots in liquid nitrogen.

5.1.1 Preparation of nuclei for transport studies

Rat liver nuclei or nuclei from cells grown in culture may be used in these extracts to study nuclear protein import. Alternatively, bacteriophage DNA or amphibian sperm pronuclei can be added to the extract. The nuclei that are assembled import nuclear proteins.

Protocol 14. Preparation of sperm pronuclei

Materials

- 1 male *Xenopus* frog
- chorionic gonadotrphin (Chorulon, Intervet Laboratories)
- SNP buffer: 250 mM sucrose; 75 mM NaCl; 0.5 mM spermidine; 0.15 mM spermine
- lysolecithin (phosphatidyl ethanolamine)

Method

1. Inject male frog with 150 IU chorionic gonadotrophin 1 week before sperm are required.
2. Remove testes and homogenize in 2 ml SNP buffer.
3. Remove debris by brief centrifugation (10 sec) in a microcentrifuge.
4. Concentrate sperm by centrifugation at 1000 *g* for 5 min.
5. Resuspend pelleted sperm in 0.5 ml SNP buffer.
6. De-membranate sperm by adding 20 μl of 1 mg/ml lysolecithin and incubate for 10 minutes at room temperature.
7. Stop the reaction by the addition of 1 ml SNP buffer containing 3% BSA.
8. Wash 3 times in SNP buffer by cycles of centrifugation and resuspension and finally resuspend in 1 ml SNP buffer containing 30% glycerol. The concentration of sperm pronuclei can be estimated using a haemocytometer or by measuring the absorption at 260 nm (A_{260}; see *Protocol 15*, footnote *a*).

Protocol 15. Preparation of rat liver nuclei

Materials

- livers from male albino rats
- Potter–Elvejhem homogenizer with motor-driven pestle (clearance 0.025 cm)
- TKM buffer: 250 mM sucrose; 50 mM Tris-HCl, pH 7.5; 25 mM KCl, 5 mM $MgCl_2$

Protocol 15. *Continued*

Method

1. Chill livers in several volumes of TKM buffer on ice.
2. Estimate volume of liver and mince in 2 volumes of TKM buffer.
3. Homogenize using 10–15 strokes of the homogenizer pestle at 1700 r.p.m.
4. Filter homogenate through four layers of cheese cloth.
5. Bring homogenate to 1.6 M in sucrose by the addition of the required volume of 2.4 M sucrose in TKM buffer.
6. Layer homogenate over 1 ml of 2.3 M sucrose in TKM buffer and centrifuge in a Beckman SW 40 rotor at 39 000 r.p.m. for 30 min (124 000 *g* average).
7. Pour off supernatant and resuspend nuclear pellet in TKM buffer.[a]

[a] The number of nuclei can be measured precisely using a haemocytometer or estimated by assuming that one A_{260} unit is equivalent to 3×10^6 nuclei. 40 grams of liver should yield about 1000–1500 A_{260} units of nuclei.

Protocol 16. *In vitro* transport assay

Materials

- *Xenopus* egg extract
- rat liver nuclei or sperm pronuclei at a concentration of $0.5–4 \times 10^5$ nuclei/μl
- fluorescently labelled or radiolabelled probe
- formaldehyde
- bis benzamide dye (Hoechst 33258; Sigma)

Method

1. To 20 μl of *Xenopus* egg extract add ATP, creatine phosphate, and creatine kinase to give final concentrations of 2 mM ATP, 9 mM creatine phosphate, and 100 U/μl creatine kinase.
2. Add 1 μl of the nuclear suspension and incubate for 30 min at 23 °C. This allows the extract to heal any breaks in the nuclear envelope. Up to 4 h incubation may be needed for nuclear assembly on naked DNA.
3. Add 1 μl of labelled probe protein at a concentration of 12–15 ng/μl.
4. Incubate at 23 °C. Normally a time course is run and accumulation is generally complete after 1 to 2 h.
5. Place 4 μl of extract on a microscope slide, add 0.5 μl 37% formaldehyde and 0.5 μl of 10 μg/ml bis benzamide dye. When a fluorescent probe is

used a coverslip is applied and accumulation is seen as protein fluorescence co-localized with the DNA fluorescence. When no accumulation occurs, non-fluorescent nuclei are seen against a fluorescent background. When radiolabelled probes are used direct autoradiography of the fixed sample on the slide is carried out using liquid photographic emulsion (Kodak).

5.2 Transport in isolated yeast nuclei

With the recent cloning of two yeast genes for nuclear pore proteins and the powerful genetics available for yeast, *in vitro* systems will prove invaluable in the biochemical characterization of yeast mutants defective in nuclear transport. In a previous volume in this series a chapter is devoted to the preparation of yeast nuclei (12). The methods described here are somewhat similar, but I have chosen to provide full protocols since these methods have been used specifically to study nuclear protein import (13). The yeast nuclei prepared by the method described here acccumulate more than 60% of the SV40 large T antigen and *Xenopus* nucleoplasmin added to the system within 15 min at 30 °C. However, it is interesting to note that, in contrast to nuclei from higher eukaryotes, transport in these nuclei is not inhibited by the addition of lectins such as wheat germ agglutinin and the nuclei show low levels of O-linked *N*-acetyglucosamine. This may represent a true difference between yeast and other nuclei or may be due to digestion by hexosaminidase contamination in the enzyme used to remove the yeast cell wall in the early steps of nuclear isolation.

Protocol 17. Growth of yeast and preparation of yeast nuclei

Materials

- YPD growth medium: 1% yeast extract; 2% bactopeptone; 2% dextran
- resuspension buffer: 10 mM Tris-SO_4, pH 9.4; 10 mM DTT; 1 mM PMSF
- lyticase (Sigma)
- β-glucuronidase from *Helix pomatia* (Sigma)
- cytochalasin B (1 mg/ml in dimethyl sulphoxide (DMSO))
- sorbitol/Hepes buffer: 1.2 M sorbitol; 20 mM Hepes, pH 7.4; 1 mM $MgCl_2$; 1 mM PMSF
- sorbitol/Pipes buffer: 1.2 M sorbitol; 20 mM Pipes, pH 6.5; 1 mM $MgCl_2$; 1 mM PMSF
- lysis buffer: 18% Ficoll; 20 mM Pipes, pH 6.5; 0.5 mM $MgCl_2$; 1 mM PMSF
- SHMCK buffer: 0.2 M sucrose; 50 mM Hepes, pH 7.4; 5 mM $MgCl_2$; 3 mM $CaCl_2$; 25 mM KCl

Protocol 17. *Continued*

Method

1. Grow a 1–2 litre culture of yeast to mid-log in YPD medium and centrifuge at 3000 *g* for 5 min. Determine the weight of yeast cells harvested.
2. Resuspend in 2 ml of resuspension buffer for every gram of packed cells.
3. Incubate at 30 °C for 15 min with gentle shaking.
4. Centrifuge at 3000 *g* for 5 min and resuspend in sorbitol/Hepes buffer using 5 ml of buffer for every 1 g of cells. Repeat and finally resuspend in 5 ml of the same buffer.
5. To resuspended cells add 2 mg of lyticase and 0.3 mg of β-glucuronidase for every 1 g of cells. Incubate at 30 °C until spheroplasts form.
6. Stop by the slow addition of 10 ml of cold sorbitol/Pipes buffer for every 1 g of cells. Incubate on ice for 5 min.
7. Centrifuge at 3000 *g* for 5 min and resuspend spheroplasts in 20 mM Hepes, pH 7.4; 5 mM $MgCl_2$; 1 mM PMSF. Mix gently at 4 °C for 10 min.
8. Add cytochloasin B to a final concentration of 10 μg/ml and incubate in the dark for 10 min at 4 °C.
9. Layer the incubation mixture (15–18 ml) over a glycerol/Ficoll step gradient in Beckman SW30 tubes.[a]
10. Centrifuge at 15 000 *g* for 40 min at 4 °C. A crude nuclear pellet is produced.
11. Resuspend pellet in 5 ml of lysis buffer and clear by centrifugation at 7000 *g* for 5 min at 4 °C.
12. Recover supernatant and centrifuge at 11 000 *g* for 20 min at 4 °C.
13. Repeat steps **10** and **11**.
14. The pure nuclear pellet, representing 50% of the starting DNA, is resuspended in SHMCK buffer to a protein concentration of 1.25 mg/ml.[b]

[a] Each gradient consists of 5 ml 40% glycerol/8% Ficoll; 10 ml 30% glycerol/8% Ficoll; 5 ml 20% glycerol/8% Ficoll in 20 mM Pipes, pH 6.5; 0.5 mM $MgCl_2$; 1 mM PMSF.

[b] These nuclei have an intact nuclear envelope and are free of cytoplasmic contamination.

Protocol 18. *In vitro* transport assay using yeast nuclei

Materials

- suspension of yeast nuclei

- the protein under investigation is produced by transcription of the cDNA *in vitro* and translation of the RNA using [^{35}S]methionine in a reticulocyte or wheat germ *in vitro* translation system [14]

Method

1. Add 25 μg of nuclear suspension (20 μl at 1.25 mg/ml) to 25 μl of SHMCK buffer (*Protocol 17*).
2. Add 3 μl of the *in vitro* translation reaction mixture containing the labelled protein.
3. Incubate at 30 °C for an appropriate time. Accumulation normally reaches a maximum after 30–60 min incubation
4. Isolate nuclei for subsequent analysis of protein uptake by layering over 50 μl of 35% sucrose; 50 mM Hepes, pH 7.4; 5 mM $MgCl_2$ and centrifuge for 1 min in a bench top microcentrifuge.

Using a slighly different protocol for the isolation of yeast nuclei the dependence of protein import on ATP hydrolysis has been demonstrated (10). This system utilizes the yeast nuclear proteins Mcm 1 and Ste 12 synthesized by *in vitro* transcription and translation. Radiolabelled BSA, carbonic anhydrase, and *β*-lactoglobulin were used as non-nuclear controls. The significant change from *Protocol 18* is the use of two protease inhibitor mixtures: one contains 5 mg/ml pepstatin A, 1 mg/ml chymostatin, and 17.4 mg/ml PMSF in dimethyl sulphoxide (DMSO), while the second contains 5 mg/ml aprotinin, 131 mg/ml aminocaproic acid, and 208 mg/ml *p*-aminobenzamidine in water. After spheroplast formation these mixtures are added to a 1/1000 dilution. Nuclei are pelleted on to a 2 M sucrose cushion to minimize damage.

The nuclear transport assay buffer system is 50 mM Hepes/KOH, pH 7.4; 250 mM sucrose; 5 mM $MgSO_4$; 3 mM $CaCl_2$; 25 mM potassium acetate; 5 mM ATP; 2 mg/ml BSA; 0.4 mg/ml cycloheximide. A 50 μl reaction mixture containing 20 μl of the nuclear suspension at 10 mg/ml nuclear protein ($\sim 3 \times 10^7$ nuclei) and 5 μl of translation mix are incubated as required and then layered on to 100 ml of cold 30% sucrose in assay buffer in a microcentrifuge tube. Nuclei are then pelleted by centrifugation at 9000 *g* for 3 min at 4 °C. The fractions are assayed by gel electrophoresis and autoradiography.

5.3 Transport in isolated rat liver nuclei

Specific, efficient nuclear uptake of SV40 large T antigen into isolated rat liver nuclei has been reported (5). The specificity of the uptake was demonstrated by the failure of the SV40 mutant sequence (in which lysine residue 128 is changed to a threonine or asparagine) to accumulate and its inability to compete with the wild-type sequence for uptake. The nuclei are prepared using the standard protocol given in *Protocol 13* and must be used the same day as they are prepared.

Protocol 19. *In vitro* transport assay using isolated rat liver nuclei

Materials

- rat liver nuclei at 2×10^5 nuclei/μl as detailed in *Protocol 13*.
- radiolabelled or fluorescently labelled probe protein
- STMKC buffer: 250 mM sucrose; 50 mM Tris-HCl, pH 7.4; 25 mM KCl; 5 mM $MgCl_2$; 3 mM $CaCl_2$
- buffer A: 50 mM Tris-HCl, pH 7.4; 25 mM KCl; 2.5 mM $MgCl_2$; 3.3 mM $CaCl_2$; 5 mM Na_2HPO_4; 5 mM spermidine

Method

1. Wash nuclei in STMKC buffer by one cycle of centrifugation and resuspension to required final nuclear concentration.
2. Add 5 μl nuclei and up to 5 ng of labelled protein to 45 μl of buffer A.
3. Incubate at 20 °C for the required time. With the SV40 nuclear targeting sequence accumulation is complete within 30–45 min.
4. Isolate nuclei by layering the incubation mixture on to 150 μl of 35% sucrose; 50 mM Tris-HCl, pH 7.4; 5 mM $MgCl_2$ in a 250 μl microcentrifuge tube.
5. Centrifuge for 30 sec.
6. Remove supernatant and resuspend nuclei in buffer for SDS gel electrophoresis. Alternatively, when using fluorescent probes the reaction mixture can be fixed in by the addition of formaldehyde and the reaction mixture is viewed as for nuclei incubated in *Xenopus* egg extracts (*Protocol 16*).

5.4 Transport in nuclear envelope vesicles

Closed nuclear envelope vesicles can be produced by heparin treatment of isolated nuclei (6). These vesicles have an apparently intact nuclear envelope and are capable of specific import of histones and *Xenopus* nucleoplasmin; non-nuclear proteins such as myoglobin and immunoglobulin are not imported. The import is rapid and efficient; more than 90% of the imported protein is taken up within 5 min.

The vesicles can be loaded with RNA during the heparin extraction step and have been used to demonstrate that the efflux of poly A^+RNA is enhanced by the presence of 2 mM ATP, while the efflux of ribosomal RNA is not.

Protocol 20. Nuclear envelope vesicle preparation

Materials

- rat liver nuclei prepared using the method described in *Protocol 15*
- heparin
- wash buffer: 250 mM sucrose; 50 mM Tris-HCl, pH 7.4; 25 mM KCl; 5 mM $MgCl_2$; 3.3 mM $CaCl_2$
- vesicle resuspension buffer: 50 mM Tris-HCl, pH 7.4; 25 mM KCl; 5 mM $MgCl_2$; 0.5 mM $CaCl_2$; 5 mM NaCl; 2.5 mM Na_2HPO_4; 5 mM spermidine

Method

1. Resuspend nuclei in 10 mM Tris, 10 mM Na_2HPO_4, pH 8.0 to give a final DNA concentration of 200 μg/ml using the A_{260} value; see *Protocol 15*).
2. Add heparin to give a 1.5-fold weight excess over the DNA, i.e. 300 μg/ml.
3. Stir for 4 min at 22 °C and filter through scrubbed nylon fibre (Associated Biomedic Systems Inc. type 200).
4. Centrifuge at 5000 *g* for 10 min at 4 °C and resuspend in wash buffer.
5. Resuspend vesicles in vesicle resuspension buffer to a final protein concentration of 2 mg/ml.[a]

[a] Vesicles range in size from 4 to 9.2 μm, 72% of the vesicle population fall in the 6.5–7.5 μm range (nuclei are 7.5 μm in diameter).

Protocol 21. *In vitro* transport assay using nuclear envelope vesicles

Materials

- nuclear envelope vesicles
- radiolabelled or fluorescently labelled probe protein

Method

1. Prepare microcentrifuge tubes (Beckman) containing 10 μl of 60% perchloric acid overlayered with 40 μl of silicone oil.
2. Prepare transport reactions containing 150 μl of nuclear envelope vesicles in vesicle resuspension buffer described in *Protocol 20* and labelled protein at a final concentration of 100 μg/ml.
3. After incubation at 22 °C layer the transport reaction on the silicone oil in the microcentrifuge tube.

Protocol 21. *Continued*

4. Centrifuge in a microcentrifuge for 1 min.
5. Remove supernatant and analyse supernatant and pelleted vesicles by gel electrophoresis and autoradiography or scintillation counting. Fluorescent proteins can be used in this assay and accumulation is observed by direct observation of the vesicle suspension with or without fixation using a fluorescence microscope.

5.5 Transport in permeabilized whole cells

In this method a cytosolic extract is prepared and added to the permeabilized cells (8). The results obtained with this system indicate that cytoplasmic proteins that are sensitive to treatment with *N*-ethylmaleimide are required for efficient nuclear import.

Protocol 22. Preparation of cytosol fractions

Materials

- HeLa or HTC (rat hepatoma cells) cells growing exponentially in suspension
- wash buffer; 10 mM Hepes, pH 7.3; 110 mM potassium acetate; 2 mM magnesium acetate; 2 mM DTT
- lysis buffer: 5 mM Hepes, pH 7.3; 10 mM potassium acetate; 2 mM magnesium acetate; 2 mM DTT; 20 μg/ml cytochalasin B; 1 mM PMSF; 1 mg/ml aprotinin, leupeptin, and pepstatin
- transport buffer: 20 mM Hepes, pH 7.3; 110 mM potassium acetate; 5 mM sodium acetate; 2 mM magnesium acetate; 1 mM EGTA; 2 mM DTT; 1 μg/ml aprotinin, leupeptin, and pepstatin

Method

1. Collect cells by low-speed centrifugation (1000 *g* for 15 min) and wash twice with cold phosphate-buffered saline (PBS), pH 7.4 by resuspension and centrifugation.
2. Wash cells in wash buffer and resuspend pelleted cells in 1.5 volumes of lysis buffer. Allow to swell for 10 min on ice.
3. Lyse cells by five strokes of a Dounce homogenizer with a tight-fitting pestle.
4. Centrifuge homogenate at 1500 *g* for 15 min to remove nuclei and cell debris.
5. Centrifuge supernatant at 15 000 *g* for 20 min, collect supernatant, and centrifuge at 100 000 *g* for 30 min.

6. Dialyse 100 000 *g* supernatant against transport buffer using a 25 000 molecular weight cut off dialysis membrane.[a]

[a] Protein concentration should be determined and should fall in the 50–60 mg/ml range. The extract can be frozen in small aliquots in liquid nitrogen and stored at −80 °C.

Protocol 23. Cell permeabilization and *in vitro* transport

Materials

- HeLa or normal rat kidney (NRK) cells grown on coverslips to 50–70% confluence
- cytosol fraction
- transport buffer (*Protocol 22*)
- 20 mg/ml digitonin (Sigma) in DMSO
- fluorescently labelled transport probe

Method

1. Prepare a complete transport mix containing 60–70% cytosol in transport buffer plus 100 nM transport probe, 5 mM creatine phosphate, and 20 U/ml creatine phosphokinase.
2. Grow cells on coverslips to 50–70% confluence.
3. Rinse coverslips with ice-cold transport buffer containing 40 μg/ml digitonin and allow to permeabilize for 5 min. Rinse with ice-cold transport buffer.
4. Blot to remove excess buffer and invert on to a drop of complete transport mix.
5. Incubate in a humidified box at 30 °C. Transport can be assayed by fluorescence by rinsing the coverslip in transport buffer, mounting on a slide in a small volume of transport buffer, and sealing the edge with nail polish.

References

1. Danscher, G. (1981). *Histochemistry* **71**, 81.
2. Dworetzky, S. I. and Feldherr, C. M. (1988). *J. Cell Biol.* **106**, 575.
3. Lanford, R. E., Kanda, P., and Kennedy, R. C. (1986). *Cell* **46**, 575.
4. Chelsky, D., Ralph, R., and Jonak, G. (1989). *Mol. Cell Biol.* **9**, 2487.
5. Markland, W., Smith, A. E., and Roberts, B. L. (1987). *Mol. Cell Biol.* **7**, 4255.
6. Reidel, N., Bachmann, M., Prochnow, D., Richter, H. -P., and Fasold, H. (1987). *Proc. nat. Acad. Sci., USA* **84**, 3540.

7. Newmeyer, D. D., Finlay, D. R. and Forbes, D. J. (1986). *J. Cell Biol.* **103**, 2091.
8. Adam, S. A., Marr, R. S., and Gerace, L. (1990). *J. Cell Biol.* **111**, 807.
9. Parnaik, V. K. and Kennady, P. K. (1990), *Mol. Cell Biol.* **10**, 1287.
10. Garcia-Bustos, J. F., Wagner, P. and Hall, M. N. (1991). *Exp. Cell Res.* **192**, 213.
11. Lohka, M. J. and Masui, Y. (1984). *J. Cell Biol.* **98**, 1222.
12. Lohr, D. (1988). In *Yeast: a practical approach* (ed. I. Campbell and J. H. Duffus). p. 25. IRL Press, Oxford.
13. Kalinich, J. F. and Douglas, M. G. (1989). *J. Biol. Chem.* **264**, 17979.
14. Jackson, R. J. and Hunt, T. (1983). *Methods in Enzymology* **96**, 50–74.

8

In vitro reconstitution of endocytic vesicle fusion

JEAN GRUENBERG and JEAN-PIERRE GORVEL

1. Introduction

It is now well established that, after internalization by clathrin-coated pits, solutes and receptors first appear in early endosomes at the cell periphery, then in late endosomes in the perinuclear region, and eventually in lysosomes (reviews in references 1–3). We have observed that tracers leaving the early endosomes appear in large spherical vesicles before they reach late endosomes (4). These large vesicles resemble one of the subpopulations of multivesicular endosomes described by Dunn *et al.* (5). Tracers internalized after microtubule depolymerization pass through early endosomes and reach these large vesicles but they do not appear in the late endosomes. We thus proposed that these vesicles are involved in the microtubule-dependent transport between early and late endosomes and may correspond to the vesicles observed to move *in vivo* between the cell periphery and the perinuclear region. In this chapter we will refer to these vesicles as endosomal carrier vesicles.

Progress has been made in understanding some of the mechanisms that are involved in the regulation of endocytic membrane traffic by using cell-free assays (reviewed in reference 2), an approach pioneered by Rothman and his collaborators in studies of transport within the Golgi complex (see references in reference 6). In this chapter we will discuss the experimental approach we have used to study fusion events within the endosomal apparatus. Membrane traffic between endosomes and other subcellular compartments has also been studied *in vitro*, including the formation of coated vesicles at the plasma membrane (7), the recycling of transferrin back to the cell surface (8), and the transfer from endosomes either to the trans-Golgi network (9) or to lysosomes (10).

Several studies have shown that early endocytic vesicles are fusogenic *in vitro* (11–17). Both organellar partners of this fusion reaction correspond to early endosomal elements and the efficiency of the process is ≈ 60% when measured with immunoisolated fractions (4). These observations suggest that individual early endosomal elements are connected by lateral interactions *in vivo*, forming a highly dynamic network that, in effect, corresponds to a single functional compartment. Some of the mechanisms involved in the control of this early endocytic fusion event are becoming clearer. The process is highly specific both in non-polarized and in polarized cells (4, 18), suggesting the involvement of specific recognition sites on the cytoplasmic surface of early endosomes. Our data also suggest that this fusion event may be arrested during mitosis, in a process mediated by the cell cycle control protein kinase cdc2 (19). It is attractive to speculate that the cdc2 kinase may also mediate the observed inhibition of other steps of membrane traffic in mitotic cells (20). Finally, fusion of early endosomes is inhibited by low concentration of GTPγS (19, 21), suggesting that GTP-binding proteins are involved in the process. In fact, we have now shown that the small molecular weight GTP-binding protein rab5 is involved in the regulation of early endosome recognition and/or fusion (22). Whereas the restricted subcellular localization of rab5 (23) argues against its involvement in other pathways of membrane traffic, different factors may be common to other pathways. In particular, one factor required for transport within the Golgi (24) is also necessary for transport from the endoplasmic reticulum (ER) to the Golgi (25) and for early endosome fusion (26). Altogether, these observations suggest that some factors may be part of a common machinery and are recruited during fusion events occurring at several steps of membrane traffic, whereas other components may provide the specificity of membrane–membrane interactions observed *in vivo* and *in vitro*.

Using polarized Madin–Darby canine kidney (MDCK) cells we have recently reconstituted *in vitro* a second fusion event, which occurs at a later stage of the endocytic pathway (18). Our data suggest that this fusion event reflects the delivery of endosomal carrier vesicles to late endosomes. However, the precise sequence of events which occur in this process is not established yet. The appearance of the markers in late endosomes in the cell-free assay was facilitated by the presence of polymerized microtubules, as is the case *in vivo*. In addition, microtubule-associated proteins, in particular the mechanochemical ATPases, which mediate the movement of vesicles on microtubules *in vitro* (reviewed in reference 27), were required in the assay (18). In this chapter, we will describe the basic protocol, which we have used to study both endosomal fusion events, the lateral fusion between early endosomes and the vectorial passage between carrier vesicles and late endosomes.

2. General considerations

2.1 Cellular materials in cell-free assays

Intracellular events and components, which would otherwise be difficult to study, have been experimentally manipulated with cell-free assays. Different techniques have been used to gain access to the intracellular milieu in studies of endocytic membrane traffic. In some studies (7–9), the plasma membrane has been mechanically perforated while retaining the overall cellular organization. These perforated or semi-intact cells are advantageous when studying events which may require the maintenance of some cellular organization, for example, transport to and from the plasma membrane (7, 8). Alternatively, cells have been homogenized and then crude (11, 14–16) or more purified fractions (4, 12, 13, 17, 22) have been prepared. In these assays, vesicles or compartments are dispersed in a suspension and are free to interact with each other. In situations where they can be used, crude fractions have proven to be extremely useful because they can be easily and rapidly prepared. More purified fractions, however, make it possible to carry out a biochemical analysis of the organellar partners under scrutiny. Here, we describe a flotation gradient which separates early from late endosomes (22). We also describe our use of immunoisolation to purify endosomal fractions with a foreign antigen, the G-protein of vesicular stomatitis virus (VSV) which is first implanted into the plasma membrane and then internalized (reviewed in reference 28).

2.2 Principles governing the choice of assay and markers

In cell-free assays, the occurrence of fusion has been most commonly detected by the formation of a product resulting from a reaction between two substrates which were originally present in separate populations of vesicles (see reference 2 and outline in *Figure 1*). In most assays, advantage has been taken of the fact that the substrates of the fusion reaction can be internalized into endocytic vesicles by fluid-phase or receptor-mediated endocytosis *in vivo*. General markers as well as markers of more specialized pathways have been used. In our studies, we have selected general markers of endosomes to compare the overall fusion activity of different populations of endosomal vesicles with the same assay. These include markers internalized by fluid phase endocytosis or after binding to cell surface proteins. We have made use of the high binding affinity and low dissociation constant of avidin for biotin to provide a fusion-specific reaction, an assay analogous to that originally used by Braell (14).

2.2.1 Fluid phase markers

For fluid phase endocytosis, avidin (mol. wt. 68 kd) and biotinylated horseradish peroxidase (bHRP, mol. wt. 40 kd) provide simple markers which distribute within the endosomal content upon internalization. After the internalization step, the cells are homogenized and fractions are prepared, which are combined in the assay in the presence of cytosol and ATP. If fusion occurs, a complex is formed between avidin and bHRP. At the end of the assay, the reaction mixture is extracted in detergents in the presence of excess biotinylated insulin, as a quenching agent. The avidin-bHRP complex is then immunoprecipitated with anti-avidin antibodies and the enzymatic activity of bHRP is quantified. Since each marker originally distributes as a solute within the vesicle lumen, the amount of avidin–bHRP detected in the assay measures the extent of content-mixing between labelled vesicles undergoing fusion.

2.2.2 Covalent membrane markers

To provide a general membrane-attached marker, cell surface proteins are biotinylated and then avidin is bound to the biotinylated cell surface proteins. These complexes are then efficiently internalized and transported towards lysosomes, providing a marker of the endosomal membrane. Avidin-labelled fractions prepared from these cells are used in combination with bHRP-labelled fractions which are obtained as described in Section 2.2.1. The experimental conditions for subcellular fractionation and the cell-free assay of fusion, including quantification of avidin-bHRP complex, are the same for membrane-attached (Section 2.2.2) or fluid-phase markers (Section 2.2.1).

2.3 Labelling of the different endosomal compartments

The conditions of internalization required to label the different endosomal compartments may vary between cell-types. The conditions we use in baby hamster kidney (BHK) cells are summarized in *Table 1*. We find that 5 min after incubation at 37 °C, membrane proteins and fluid phase tracers distribute predominantly within early endosomes which exhibit a high fusion activity *in vitro* (4, 13). At longer times of internalization, the markers appear first in carrier vesicles and then sequentially in late endosomes and lysosomes. The markers can be accumulated in the carrier vesicles by incubating at 20 °C in the presence of microtubules (29) or at 37 °C after microtubule depolymerization with 10 μM nocodazole (4). The latter conditions result in a more efficient labelling, due to higher internalization rate at 37 °C when compared to 20 °C. In a typical experiment with fluid phase markers, the cells are incubated with avidin or bHRP for 10 min at 37 °C and then for an additional 30 min in marker-free medium to chase the markers from the early

Table 1. Conditions of internalization of the markers

	Time of internalization at 37 °C
Early endosomes	5 min
Carrier vesicles	10 min pulse[a] + 30 min chase[a] after MT depolymerization[b]
Late endosomes	10 min pulse[a] + 30 min chase[a]

[a] The conditions for pulse and chase internalization are indicated in *Protocol 2*.
[b] To depolymerize microtubules (MT) in BHK cells, the cells are pre-incubated during 1–2 h in the presence of 10 μM nocodazole and then nocodazole remains present in all solutions and media up to the internalization step.

endosomes. Late endosomes and to some extent lysosomes are labelled under the same conditions when microtubules are present.

3. Manipulation of intact cells

3.1 Cells

Monolayers of the baby hamster kidney (BHK-21) cell line are grown in Glasgow's minimum essential medium (G-MEM) supplemented with 5% fetal calf serum (FCS), 10% tryptose phosphate broth, and 2 mM glutamine in a 5% CO_2 atmosphere. We find that cells grow as an even monolayer, a critical requirement in these experiments, when seeded 14–16 h before use at a high density (4×10^4 cells/cm^2 of culture dish) in 10-cm Petri dishes. These conditions yield $\approx 1.3 \times 10^7$ cells per dish ($\approx$ 2.5–3.0 mg total protein) at the time of the experiment. This protocol has the additional advantage that 14–16 h old cells can easily be homogenized (30). The conditions required to depolymerize microtubules may vary between cell types. In BHK cells, we find that 1–2 h pre-incubation at 37 °C in the presence of 10 μM nocodazole is sufficient (4). (Nocodazole then remains present in all solutions and media up the internalization step.) No apparent depolymerization of BHK microtubules was observed after prolonged incubation at 4 °C. In contrast, the depolymerization of MDCK microtubules requires both a cold treatment and an incubation at 37 °C in the presence of nocodazole (see reference 18). Typically, 2–3 dishes (10 cm) are used per experimental condition, and 15–20 dishes for preparing stocks.

3.2 Fluid phase internalization

We use avidin and bHRP, which were separately internalized into two cell populations, as markers for the reconstitution of fusion *in vitro* (see *Figure 1*). Avidin can be obtained from commercial sources or prepared from egg white by a single affinity-chromatography step on 2-iminobiotin beads (available from Calbiochem, no. 401779; see reference 31). HRP is also commercially available and can be easily biotinylated without loss of enzymatic activity (4). *Protocol 1* outlines the preparation of relatively small amounts of bHRP (20 mg at a final concentration of 1.8 mg/ml), but the protocol can be scaled up to prepare larger stocks (use, for example, 100 mg HRP and 57 mg biotin-X-NHS in a proportionately smaller volume, e.g. 10–15 ml). Then bHRP can be stored in 50% glycerol at −20 °C and redialysed before use (as in *Protocol 1*).

Protocol 1. Biotinylation of HRP

Materials

- HRP

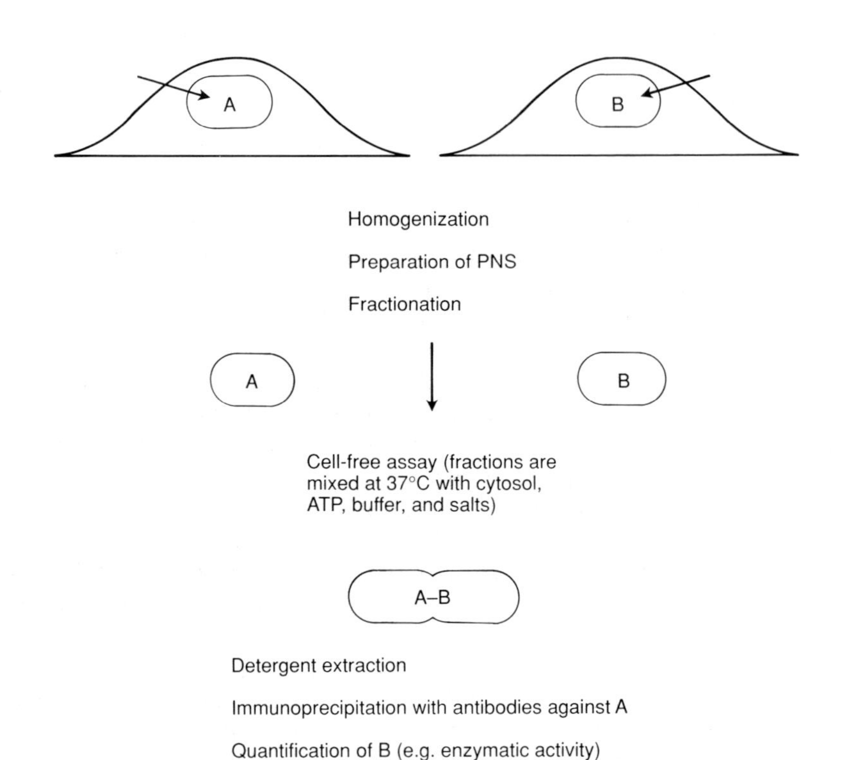

Figure 1. Outline of the cell-free assay of endocytic vesicle fusion.

Protocol 1. *Continued*

- buffer A: 0.1 M $NaHCO_3/Na_2CO_3$, pH 9.0
- biotin-X-NHS: biotinyl-ε-aminocaproic acid *N*-hydroxysuccinimide ester (Calbiochem no. 203 188)
- dimethylformamide (DMF)
- 0.2 M glycine, pH 8.0
- internalization medium: MEM; 10 mM Hepes; 5 mM D-glucose, buffered to pH 7.4

Method

1. Dissolve 20 mg HRP in 9.5 ml of buffer A. Dissolve 11.4 mg of biotin-X-NHS in 0.5 ml DMF in a glass tube. Mix and incubate with gentle stirring for 2 h at room temperature (50:1 molar excess of biotin).

2. Quench unreacted active groups with 1 ml of glycine and mix for an additional 30 min.
3. Dialyse the mixture at 4 °C against several changes of internalization medium.
4. Filter-sterilize and store at 4 °C until use.

Relatively high concentrations of avidin or bHRP must be used in the medium, in order to internalize sufficient amounts of the markers by fluid phase endocytosis. To simplify the calculation of the efficiency of vesicle fusion, equimolar concentrations of avidin and bHRP are used (3.2 mg/ml avidin and 1.8 mg/ml bHRP in the medium, respectively). The conditions of internalization of the markers are summarized in *Table 1*. Internalization is carried out directly on the Petri dish (*Protocol 2*), particularly after VSV G-protein implantation or cell surface biotinylation. After the internalization step, the cells must be thoroughly washed to remove the markers which may remain non-specifically adsorbed. Avidin- and bHRP-containing media can be re-used 3–5 times provided a total internalization time of 30 min is not exceeded. After the internalization step, the media are collected and their pH is readjusted, if necessary. They are then filtered through a 0.45-μm filter and stored at 4 °C.

When working with monolayers of non-polarized cells (like BHK), the costs of the experiment can be significantly reduced if the cells are first detached from the substratum and then resuspended in a small volume during the internalization step. The cells are released with trypsin at 4 °C and washed thoroughly by repeated centrifugations in buffer containing BSA and soybean trypsin inhibitor. While incubating cells at 37 °C and at a high density for times > 15 min, care should be taken to avoid acidification of the medium, a condition that affects endocytosis (32–7).

Protocol 2. Fluid-phase internalization of avidin and bHRP on dishes

All experimental steps and all solutions are at ice temperature except the internalization medium (and a small volume of PBS^+/BSA for the chase, if needed) which is at 37 °C.

Materials

- avidin
- bHRP
- phosphate-buffered saline$^+$(PBS^+): 137 mM NaCl; 2.7 mM KCl; 1.5 mM KH_2PO_4; 6.5 mM Na_2HPO_4; 1 mM $CaCl_2$; 1 mM $MgCl_2$

Protocol 2. *Continued*

- internalization medium (see *Protocol 1*)
- PBS^+/BSA: PBS^+ containing 5 mg/ml BSA

A. *Preparation*

1. Place 10 dishes of cells on ice plates (a wet metal plate is fitted into an ice bucket to guarantee good contact with the dish). The dish must lie perfectly flat on the plate.

2. Remove medium with an aspirator connected to a water pump, add 5 ml PBS^+ to each dish and leave for 2–3 min on a rocking platform. Repeat this step two times.

B. *Continuous internalization*

1. Aspirate the solution and pre-warm each dish for 1–2 sec in a water bath at 37 °C. To each dish add 2.5 ml of internalization medium pre-warmed at 37 °C and containing 3.2 mg/ml avidin (five dishes) *or* 1.8 mg/ml bHRP (five dishes).

2. Place dishes on a rocking platform in a 37 °C incubator (or a water bath at 37 °C) for the desired time (*Table 1*). If a chase is desired go to parts C or D; otherwise continue with part B.

3. Return the dishes to the ice-plate, aspirate the medium, and wash each dish three times for 5–10 min with 5 ml PBS^+/BSA.

4. Aspirate and wash each dish two times with 5 ml PBS^+. The dishes are ready for homogenization.

C. *Chase after internalization*

1. After the pulse internalization (part B, step **2**) return the dishes to the ice-plate and wash each dish three times for 5 min with 5 ml PBS^+/BSA.

2. Return each dish to 37 °C and further incubate at 37 °C in 10 ml internalization medium containing 2 mg/ml BSA.

3. Return the dishes to the ice-plate, aspirate the medium, and wash each dish two times for 5 min with 5 ml PBS^+/BSA.

4. Aspirate and rinse two times with PBS^+. The dishes are ready for homogenization.

D. *Alternative chase after internalization*

1. Immediately after the pulse internalization step (part B, step **2**), rinse each dish two to three times with 5 ml PBS^+/BSA (pre-warmed at 37 °C).

2. Incubate further at 37 °C in 10 ml internalization medium containing 2 mg/ml BSA.

3. Continue with steps **3** and **4** of part C.

3.3 Preparation of VSV stocks

In immunoisolation experiments, the antigen is provided by the cytoplasmic domain of the transmembrane glycoprotein G of the enveloped virus VSV. VSV is produced and harvested as described (12). Briefly, subconfluent monolayers of 2-day-old BHK cells are infected with 0.02 plaque-forming units/cell and 19–20 h later the medium, containing the virus, is collected. The cell debris is removed by centrifugation for 30 min at 7000 r.p.m. in a HS4 rotor. The supernatant is then layered on top of a two-step gradient consisting of 10 and 55% sucrose in 50 mM Tris-HCl, 100 mM NaCl, 0.5 mM EDTA, pH 7.4 and centrifuged for 90 min at 25 000 r.p.m. in a SW27 rotor. The virions are collected at the 10–55% interface, aliquoted in 50- and 200-μg aliquots, frozen in liquid nitrogen, and stored at −70 °C.

3.4 Implantation and internalization of VSV G-protein

To immunoisolate endosomal fractions, the transmembrane glycoprotein G is first introduced into the plasma membrane and then internalized (see *Protocol 3*). The G cytoplasmic domain is then exposed on the surface of endosomal elements and accessible to antibodies after homogenization. To introduce the G-protein in the plasma membrane, intact virions are first attached to the cell surface. (High and reproducible amounts of VSV can be bound to the plasma membrane via wheat germ agglutinin (WGA)). The VSV envelope is then fused with the plasma membrane by a 30-sec incubation at 37 °C in a medium at pH 4.9. At this acidic pH, endocytosis is fully inhibited (32). Low pH-mediated fusion results in the incorporation of G molecules into the plasma membrane in their correct transmembrane orientation. The remaining intact virions, which did not undergo fusion, are then removed with *N*-acetylglucosamine. Since all manipulations, except fusion itself, are at the temperature of ice, the G-protein remains restricted to the plasma membrane before the internalization step. Addition of 50 μg VSV to one 10-cm dish results in the fusion of ≈ 8.5 μg total VSV protein with the plasma membrane, which corresponds to a density of ≈ 80 implanted G molecules/μm^2 membrane surface area (12).

In our fusion experiments, the G-protein is always co-internalized with avidin present in the fluid phase, using the conditions in *Protocol 2* and *Table 1*. In the assay, the avidin-labelled fraction immobilized on the solid support is combined with a fraction prepared after bHRP internalization.

Protocol 3. Implantation of VSV G-protein into the plasma membrane

Caution. VSV is a rodent pathogen. Check with your local authorities. We sterilize all plastic and disinfect all solutions and glassware, for example with

Protocol 3. *Continued*

chloramine T. For aspiration with a vacuum pump an intermediate flask containing the disinfectant is used.

Materials

- VSV
- WGA
- 7.4-M: MEM containing 10 mM TES (*N*-tris(hydroxymethyl)methyl-2-aminoethane sulphonic acid); 10 mM MOPS; 15 mM Hepes; 2 mM NaH_2PO_4; 0.35 mg/ml $NaHCO_3$, pH 7.4
- PBS^+ (*Protocol 2*)
- PBS^-: as PBS^+ (*Protocol 2*) but without divalent cations
- fusion medium: MEM; 20 mM succinate, pH 4.9
- *N*-acetylglucosamine (GlcNAc)

Method[a]

1. Cool the Petri dish to ice temperature on an ice-plate (see *Protocol 2*, part A, step **1**). The dish must lie perfectly flat on the plate. Place the ice-plate on a rocking platform.
2. Aspirate the medium and wash two times with 5 ml PBS^+.
3. Aspirate and add to the dish 250 μg WGA in 2.5 ml 7.4-M. Incubate for 1 h.
4. Aspirate the medium and wash two times with 5 ml PBS^-.
5. Aspirate and add 50 μg VSV in 2.5 ml 7.4-M. Incubate for 1 h.
6. Aspirate and wash two times with 5 ml PBS^-.
7. Remove the dish from the ice-plate and aspirate the solution. Add 20 ml fusion medium pre-warmed to 37 °C and leave the dish for 30 sec at 37 °C.
8. Aspirate fusion medium and then rapidly place the dish on the ice-plate and rinse two times with 5 ml ice-cold PBS^-.
9. Aspirate, add 5 ml/dish 7.4-M containing 60 mM GlcNAc and leave for 30 min on ice.
10. Aspirate and wash two times with PBS^+.
11. The implanted G-protein is then co-internalized with avidin present in the fluid phase (*Protocol 2*).

[a] All solutions are ice-cold except the media for internalization and fusion, which are at 37 °C.

3.5 Biotinylation of plasma membrane proteins

To provide a fusion marker that is covalently linked to membrane proteins, the cell surface is first biotinylated with a protocol essentially similar to that described by Le Bivic *et al.* (38) as outlined in *Protocol 4* (see *Figure 1*). As a reagent, we use LC-biotin (sulphosuccinimidyl-6-(biotinamido) hexanoate; Pierce, no. 21335) or biotin-SS-NHS (sulphosuccinimidyl 2-(biotinamido) ethyl-1, 3-dithiopropionate; Pierce no. 21331). In both cases, the reactive group will spontaneously react with amino groups, allowing coupling of biotin to protein. Since the half-life of the reactive group in solution is relatively short, solutions should be freshly prepared. Alternatively, a stock solution can be prepared at 0.2 g/ml in water-free dimethylsulphoxide (DMSO) and stored at −20 °C. After biotinylation of the plasma membrane, excess avidin is bound to the biotinylated proteins. This guarantees that avidin retains free biotin-binding sites. Upon incubation at 37 °C, efficient internalization of the complexes into endosomes occurs (22; P. Schrotz and J. Gruenberg, unpublished). Conditions of internalization are as in *Protocol 2*. At the end of the 37 °C incubation, avidin remaining in the cell surface is quenched with biotinylated insulin. When biotin-SS-NHS is used, the biotin groups remaining on the cell surface can be removed by reduction of the -SS- bridge with glutathione (39). The efficiency of the cleavage reaction is ≈ 90% (P. Schrotz and J. Gruenberg, unpublished).

Protocol 4. Biotinylation of the plasma membrane proteins

Materials

- internalization medium (*Protocol 1*)
- 500 mM glycine
- 7.4-M: (*Protocol 3*)
- PBS^+ (*Protocol 2*)
- PBS^+/BSA (*Protocol 2*)
- biotin-SS-NHS
- 1 mg/ml sterile biotinylated insulin (Sigma)

Method

1. Cool the Petri dish to ice temperature on an ice-plate (see *Protocol 2*, part A, step **1**). The dish must lie perfectly flat on the plate. Place the ice-plate on a rocking platform.
2. Wash each dish three times with 5 ml ice-cold PBS^+.
3. Incubate each dish for 30 min on ice with 2.5 mg of freshly prepared biotin-SS-NHS (from a 0.2 g/ml stock solution in DMSO in 2.5 ml ice-cold PBS^+). Wash each dish with 5 ml PBS^+.

Protocol 4. *Continued*

4. Repeat step **3**.
5. Quench with 5 ml of 50 mM glycine in PBS^+ (from a 0.5 M stock) for 15 min on ice.
6. Wash each dish sequentially with:
   (a) 5 ml PBS^+;
   (b) 5 ml 7.4-M for 5 min;
   (c) 5 ml PBS^+.
7. Incubate each dish with 2.5 ml of 7.4-M containing 0.3 mg/ml avidin for 30 min on ice.
8. Wash each dish three times for 5 min with 5 ml PBS^+/BSA and then once with 5 ml PBS^+.
9. Internalization is carried out as in *Protocol 2* (part B), except that the internalization medium contains 2 mg/ml BSA (no avidin or bHRP!).
10. Quench the cell surface with 30 μg/ml biotinylated insulin in PBS^+ for 30 min on ice; repeat step **8**.
11. If needed, the chase is carried out as in *Protocol 2* (part C or D). The cells are ready for homogenization.

4. Homogenization and fractionation

4.1 Homogenization

Gentle conditions of homogenization should be used to limit possible damage to endosomal elements (30), particularly when using fluid phase markers. Clearly, the markers should remain entrapped in vesicles (latent) after homogenization (see *Protocol 6*). Harsh conditions should, however, always be avoided in order to limit the breakage of lysosomes and consequent proteolysis due to released hydrolases. Since cells grown for 14–16 h are easily homogenized, it is wise to monitor each step of the homogenization process under phase contrast microscopy. First, the cells are released from the dish by scraping with the sharp edge of a rubber policeman. At this step, cell breakage should not exceed 5–10% of the cells. After collecting the cells by centrifugation, the cells are resuspended in a small volume. Homogenization is easier at a relatively high density of cells, typically 20–30% (v/v). The cells are then homogenized by passage through a needle or the tip of a pipette and then a post-nuclear supernatant (PNS) is prepared (*Protocol 5*). Under gentle conditions of homogenization, 50–60% of a fluid phase marker is recovered in the PNS. The rest, which consists partially of unbroken cells, is lost to the nuclear pellet.

Protocol 5. Homogenization of attached cells

Materials

- PBS^+ (*Protocol 2*)
- PBS^- (*Protocol 3*)
- homogenization buffer (HB): 250 mM sucrose; 3 mM imidazole, pH 7.4 (other buffers can be 10 mM Hepes or Tris at the same pH)
- cocktail of protease inhibitors[a]
- refrigerated bench centrifuge at 4 °C

Method

1. Remove the cells from five dishes by scraping with a rubber policeman in 2.5 ml PBS^- at 4 °C.
2. Centrifuge cells at 750 r.p.m. for 5 min; analyse supernatant for the presence of markers due to cell breakage during scraping.
3. Resuspend cells very gently in 5 ml HB using a pipette with a wide tip and recentrifuge at 2500 r.p.m. for 10 min.
4. Homogenization. Add 0.5 ml HB containing protease inhibitors to the cell pellet and resuspend:
   (a) first, with a blue tip;
   (b) then, 3–10 times through a 22G needle fitted to a 1-ml syringe.

   Monitor by phase contrast microscopy. Nuclei must appear clean of cellular materials and intact. Be careful not to over-homogenize.
5. Centrifuge the homogenate at 2500 r.p.m. for 15 min and collect the PNS and the nuclear pellet. Keep a 50-µl aliquot of the PNS and the nuclear pellet (resuspended in 1 ml PBS^+) for analysis. The PNS is ready for fractionation.

[a] In cell-free assays of fusion, HB contains 10 µg/ml aprotinin, 1 µg/ml pepstatin, and 1 µg/ml antipain (from 1000 × concentrated stocks of aprotinin in H_2O and antipain and pepstatin in DMSO). See reference 12 for a more complete cocktail.

When working with cells in suspension, e.g. after trypsin treatment, homogenization may require harsher conditions. The protocol then remains essentially the same, except that a tight-fitting glass–glass Potter or a Dounce homogenizer is used. Up to 15–20 passages of the pestle may be required to achieve sufficient cell breakage.

It is essential to quantify the latency of fluid phase markers in the PNS (*Protocol 6*). Our observations indicate that a latency of $\geq$ 70% is required for the cell-free analysis. (Avidin and bHRP are quantified as in *Protocols 8* and *9*, respectively.) In all fractionation experiments, a balance sheet should be established for the distribution of protein and markers (e.g. bHRP) in all fractions (see the example in *Table 2*). This provides the only appropriate means to judge the homogenization/fractionation steps and to compare different preparations (see reference 40).

Protocol 6. Latency measurement

Materials

- HB (*Protocol 5*)

Method

1. Load a 20-μl aliquot of the PNS (see *Protocol 5*) into an airfuge tube and fill the tube with HB while mixing with the PNS.
2. Centrifuge at 4 °C for 20 min at 20 p.s.i. in an airfuge.
3. Collect the supernatant and leave the pellet in the tube (it may not be visible).
4. Determine the amount of marker in the original suspension, in the pellet, and in the supernatant.
5. Express the amount of marker in the pellet (intravesicular) as a percentage of the total amount present.

4.2 Determination of avidin and bHRP

The assay used to measure internalized bHRP in PNS, nuclear pellets, and fractions is the same as that used to quantify bHRP complexed to avidin after *in vitro* fusion. The substrates of the peroxidase are *o*-dianisidine and H_2O_2 and the brown-coloured product is quantified in a spectrophotometer at 455 nm (*Protocol 7*). Note that *o*-dianisidine should be handled with care since it is a potential carcinogen. With small amounts of HRP the enzymatic reaction can be developed for up to 2 h in the dark. In fact, when the colour develops slowly, the reaction does not need to be stopped. The prosthetic haem group of the HRP molecule can, however, be blocked at the desired time with respiratory inhibitors, e.g. 10^{-5} M KCN. The reagent (*Protocol 7*, step **1**) can be stored at 4 °C, if protected from light, as long as it does not develop a straw color.

Protocol 7. bHRP determination

Materials

- *o*-dianisidine
- 0.3% H_2O_2

Table 2. Example of balance sheet (separation of early and late endosomal fractions on the flotation gradient)[a]

	Volume of fraction (ml)	HRP (OD_{455})	Protein (mg)	HRP specific activity OD_{455}/mg protein	Yield from the homogenate (%)	RSA[b]
Early endosomes (5 min at 37 °C)						
Homogenate	0.7	4.5	11.3	0.4	100	1.0
PNS	0.6	3.0	7.37	0.4	67	1.0
Early fraction	0.4	0.6	0.15	4.0	14.7	10.0
Late fraction	0.3	0.06	0.12	0.4	1.2	1.2
Late endosomes (5 + 30 min at 37 °C)						
Homogenate	0.7	2.7	10.8	0.3	100	1.0
PNS	0.6	1.6	7.2	0.2	58	0.9
Early fraction	0.5	0.04	0.09	0.4	1.3	1.6
Late fraction	0.6	0.6	0.09	6.9	25.0	27

[a] Early endosomes were labelled after incubation in the presence of HRP for 5 min (*Protocol 2*). Late endosomes were labelled after incubation in the presence of HRP for 5 min followed by a 30-min chase in marker-free medium (*Protocol 2*). Five dishes were used for each of these two typical experiments and homogenized in 500 μl of homogenization buffer (*Protocol 5*). PNS were then prepared and loaded at the bottom of the flotation gradient (*Protocol 9*). After the run the fractions which were collected correspond to the position where the early ('Early fraction') and late ('Late fraction') endosomes are recovered. Note that the yield is reduced by the loss of a significant fraction of the marker to the nuclear pellet under our gentle conditions of homogenization.

[b] RSA, Relative specific activity, the specific activity in the fraction divided by specific activity in the homogenate.

Protocol 7. *Continued*

- 0.5 M Na phosphate buffer, pH 5.0
- 2% TritonX-100
- 1.0 mM KCN
- bHRP

Method

1. Prepare reagent as follows. Using very clean glassware (e.g. as used for tissue culture), mix 12 ml of 0.5 M Na phosphate buffer, pH 5.0 and 6.0 ml of 2% Triton X-100 with 100.8 ml of double distilled H_2O. Add 13 mg *o*-dianisidine, dissolve gently, and add 1.2 ml of 0.3% H_2O_2 (total volume = 120 ml). Avoid magnetic stirring.
2. Prepare samples, blanks, and standards containing 1–10 ng of HRP, each in 0.1 ml of the same buffer as present in the samples.
3. Add 0.9 ml of the reagent to each tube, mix quickly, and record the time with a stop clock.
4. When a brown colour has developed read the absorbance at 455 nm (A_{455}) and record the time. Results are expressed as OD (optical density) units/min or ng(HRP)/min. Stop the reaction with 10 μl of 1.0 mM KCN, if necessary.

To quantify avidin, it is important to use the same detection system as used in the *in vitro* assay (*Protocol 8*). The fraction is solubilized in detergent and saturating amounts of bHRP are added. (It is advisable to titrate the amount of bHRP required for saturation.) Avidin complexed to bHRP is then immunoprecipitated with anti-avidin antibodies. We use 10 μg of our affinity-purified antibody immobilized to Protein A–Sepharose for the immunoprecipitation of $\leq$ 1 μg of avidin with an efficiency $>$ 80%. Affinity-purified antibodies against avidin are commercially available (for instance, Calbiochem no. 189729). However, the conditions of quantitative immunoprecipitation should be established for each antibody–antigen couple and routinely checked. After immunoprecipitation, the enzymatic activity of the complexed bHRP is quantified (*Protocol 7*).

Other protocols can also be used. For instance, internalized avidin can easily be monitored after its radioiodination or with an ELISA (4).

Protocol 8. Detection of avidin

Materials

- PBS^- (see *Protocol 3*)

- PBS⁻/BSA: PBS⁻ containing 5 mg/ml BSA
- Protein A–Sepharose slurry as a 50% suspension of the beads in PBS⁻ (Pharmacia)
- 10% TritonX-100 in PBS⁻
- anti-avidin antibody (affinity purified)

Method

1. Prepare beads as follows. Use 50 μl of Protein A–Sepharose slurry per experimental point. For 10 points:
   (a) Centrifuge 500 μl of slurry at 3000 r.p.m. for 2–3 min in a microcentrifuge, aspirate the solution, resuspend in 1 ml PBS⁻/BSA, and recentrifuge. Repeat this washing step five times.
   (b) Resuspend the slurry to a final volume of 1 ml with PBS⁻/BSA and add 100 μg of affinity-purified antibody against avidin.
   (c) The mixture is rotated end-over-end overnight at 4 °C.
   (d) Wash the beads five times in PBS⁻/BSA by centrifugation as in step **1(a)**, and resuspend in a final volume = 500 μl with PBS⁻/BSA.
2. Solubilize the desired fraction for 30 min at 4 °C after complementing with Triton X-100 to 0.5% (e.g. add 5 μl of 10% Triton X-100 to a 100-μl sample) and then dilute to 1 ml with PBS⁻/BSA.
3. Add 50 μl of Protein A–Sepharose slurry[a] with bound anti-avidin antibodies and rotate end-over-end overnight at 4 °C.
4. Wash the beads five times by centrifugation as in step **1(a)** but using 1 ml PBS⁻/BSA containing 0.2% Triton X-100. Wash one more time as in step **1(a)** but using 1 ml PBS⁻.
5. Resuspend the beads in 100 μl PBS⁻ and measure HRP using *Protocol 7*.

[a] To aid pipetting of the slurry, cut the last 3–4 mm off the end of a yellow pipette tip.

4.3 Flotation gradient

Endosomal fractions can easily be prepared by flotation in a sucrose/D_2O gradient (*Protocol 9*). The PNS is loaded in 40.6% sucrose and overlaid successively with 16% sucrose in D_2O, with 10% sucrose in D_2O, and eventually with homogenization buffer. With this gradient, rab5-positive early endosomes can be separated from late endosomes, which contain rab7 and the cation-independent mannose 6-phosphate receptor (22). In this latter study, we used the gradient to investigate the role of rab5 in the fusion of early endosomes *in vitro*. Our more recent studies indicate that the endosomal carrier vesicles are recovered at the same position as late endosomes on the gradient. We have now used this gradient to reconstitute the fusion between carrier vesicles and late endosomes *in vitro* (F. Aniento, N. Emans, and J. Gruenberg, unpublished; see also reference 18). The gradient is outlined in *Figure 2*.

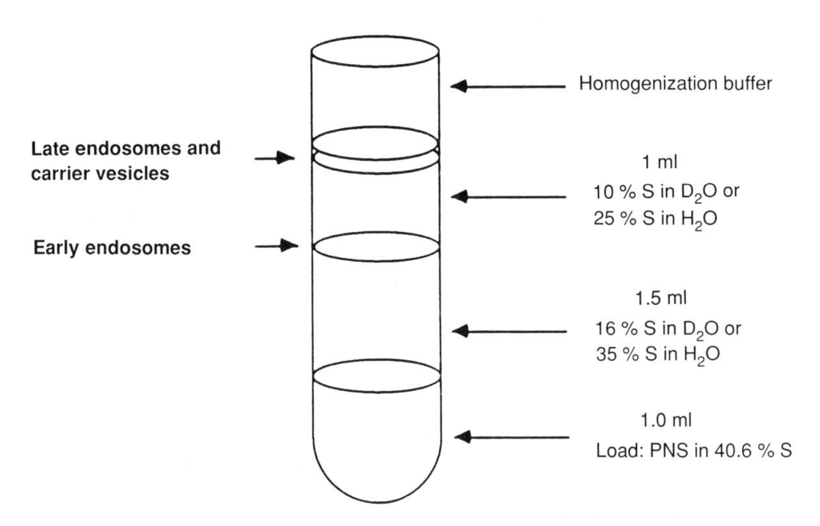

Figure 2. Flotation gradient. The PNS is brought to 40.6% sucrose (S) and loaded at the bottom of an SW60 tube. The load is then overlaid sequentially with 16% sucrose in heavy water (or 35% sucrose), 10% sucrose in heavy water (or 25% sucrose), and finally with homogenization buffer. This gradient is run for 60 min at 35 000 r.p.m. Early endosomes and late endosomes (+ carrier vesicles) are collected as indicated.

Alternatively, fusion-competent early and late endosomal fractions can also be prepared in a similar gradient, consisting of 35% and 25% sucrose steps in H_2O using the same buffer (*Protocol 9*) instead of 16% and 10% sucrose in D_2O, respectively. Finally, the low-density sucrose step can be omitted altogether, yielding fractions containing mixed populations of early and late endosomes. These fractions can be used to reconstitute both early and late endosomal fusion events (18, 19).

Protocol 9. Preparation of endosomal fractions in a flotation gradient

Materials

- 62% sucrose: 62% sucrose (w/w) in 3 mM imidazole, pH 7.4 containing 1 mM EDTA
- 16% sucrose: 16% sucrose (w/w) in D_2O, 3 mM imidazole, pH 7.4 containing 0.5 mM EDTA

- 10% sucrose: 10% sucrose (w/w) in D_2O, 3 mM imidazole, pH 7.4 containing 0.5 mM EDTA
- HB (see *Protocol 5*)

Method

1. Adjust the PNS to 40.6% sucrose and 0.5 mM EDTA using 62% sucrose (≈ 1:1 dilution of the PNS).
2. Load 1 ml of the PNS in 40.6% sucrose at the bottom of an SW60 tube and overlay sequentially with 1.5 ml 16% sucrose and then with 1 ml 10% sucrose. Fill with HB.
3. Mount the tubes in an SW60 rotor; spin for 1 h at 35 000 r.p.m. at 4 °C.
4. Collect early and late endosomal fractions from the 16/10% interface and the upper portion of the 10% cushion, respectively (see *Figure 2*). Work at 4 °C and use a peristaltic pump fitted to a 50-μl capillary. Also collect the rest of the gradient for analysis (balance sheet).
5. Aliquot early and late endosomal fractions (50 μl), freeze, and store in liquid nitrogen. The fractions are ready to be used in the cell-free assay.

4.4 Immunoisolation

We have previously described in detail the technique we use for immunoisolation with the cytoplasmic domain of the G-protein as antigen (28, 30). Here, we will briefly summarize some of the important aspects of the technique (for a review see reference 41 and also see *Protocol 10*). Different solid supports are available: (a) heterodisperse polyacrylamide beads (Bio-Rad); (b) *S. aureus* expressing protein A (Pansorbin, Calbiochem); (c) monodisperse magnetic beads (Dynal; see also reference 28); (d) Eupergit particles (Röhm Pharma Gmbh); (e) cellulose fibres (not commercially available; see reference 42). The efficiency of immunoisolation for one particular antigen/antibody couple may vary with the different immunoadsorbent, depending on the abundance and/or accessibiity of the antigen on the membrane of interest. Clearly, the actual immunoisolation experiment will provide the only appropriate trial for selecting an immunoadsorbent. In our system, the G-protein of VSV is present at an estimated density of 80 G molecules/μm^2 membrane surface area (12). The cytoplasmic domain of the G-protein is relatively short (consisting of the 22 carboxyl terminal amino acids) and the antibody we have used was raised against a synthetic carboxyl terminal peptide by Kreis (43). To reconstitute fusion in the cell-free assay, we have used magnetic beads (Dynal no. M-450) with a coupled linker antibody raised in sheep against the Fc domain of mouse IgG (for preparation see reference 28) or anti-mouse polyacrylamide beads (BioRad, anti-mouse immunobeads, no. 170–5104). In our system > 70% of fluid phase markers internalized for 5 to 30 min could be immunoisolated, when adding ≤ 300 μg of PNS protein to the beads, with an enrichment of 10–15 × (12). Enrichment is calculated from the ratio of markers immunoisolated on specific beads to the amount of markers bound non-specifically to control beads.

The immunoisolated fractions immobilized on solid supports retain their fusion activity (12, 13), with only a minimal decrease in latency of fluid phase markers (4). A significant advantage is that the fraction can be introduced into and then retrieved from a reaction mixture at the end of the reaction, while bound to the solid support (4, 12, 13). Since retrieval is simple, this separation can be repeated sequentially with the same fraction through a series of experimental conditions. This is easily achieved with magnetic beads, since they can be sedimented within minutes using a small magnet.

In fusion analysis, the G-protein is implanted into the plasma membrane (*Protocol 3*) and then co-internalized with avidin (*Protocol 2*). The cells are then homogenized and a PNS is prepared (*Protocol 5*). Immunoisolation (*Protocol 10*) can be carried out directly from the PNS or from fractions collected after flotation, preferentially using the sucrose/D_2O gradient (*Protocol 9*). A flow chart of the experiment is outlined in *Figure 3*. In all cases, the specific antibody is first coupled to the beads and then the complex is added to the fractions, as direct addition of antibodies to the fraction may lead to aggregation of antigen-containing vesicles, a condition that increases non-specific adsorption. *Protocol 10* outlines the immunoisolation experiment. To prepare the immunoadsorbent, the specific antibody is added to polyacrylamide or magnetic beads with a covalently coupled linker antibody against mouse-IgG. In both cases, 1 mg of solid support (corresponding to a pellet of ≈ 20 μl) is used for immunoisolation from ≈ 300–500 μg PNS protein or 50 μg protein of a fraction collected from the gradient. These amounts correspond approximately to the equivalent of BHK cells grown on 10–20 cm^2 of tissue culture surface area. After the experiment, the amount of immunoisolated protein cannot be directly measured because of the presence of BSA and IgG. An estimate can, however be obtained when using metabolically labelled cells, e.g. with ^{35}S-Met. When using magnetic beads, the magnetic properties should only be used to retrieve the beads after the immunoisolation step itself, to limit magnetic aggregation of the beads.

Protocol 10. Immunoisolation using the cytoplasmic domain of the G-protein as antigen

Materials

- PBS^-/BSA (see *Protocol 8*)
- PBS^- (see *Protocol 3*).

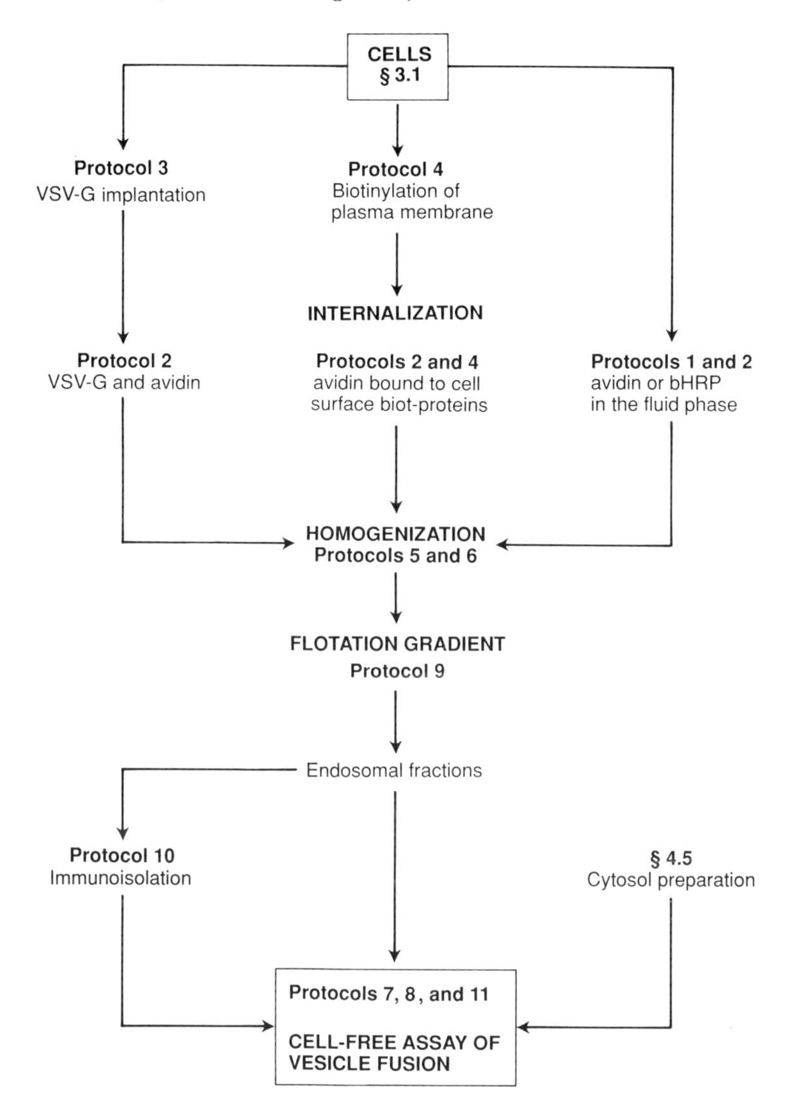

Figure 3. Flow chart of the experiment.

Method

1. To bind the specific antibody to beads:

(a) Resuspend 1–10-mg anti-mouse beads in 1 ml PBS^-/BSA and centrifuge at 3000 r.p.m. for 2–3 min in a microcentrifuge.

(b) Aspirate the supernatant, resuspend in 1 ml PBS^-/BSA and centrifuge as in (a). Repeat the washing by centrifugation three

Protocol 10. *Continued*

times. Then add 5–10 μg of specific antibody per mg bead in a total volume of 1 ml PBS⁻/BSA for 1–10 mg of beads.

(c) Rotate end-over-end overnight at 4 °C.

(d) Remove the excess antibody by repeating 3 times the washes by centrifugation as in (b) and resuspend 1 mg of beads in 100 μl PBS⁻/BSA.

2. To prepare the cellular fraction dilute the desired fraction three times or more with ice-cold PBS⁻/BSA in a microcentrifuge tube (to a final volume ≈ 1 ml).

3. Immunoisolation. Add 1 mg beads with bound antibodies to the cellular fraction and rotate end-over-end for 2 h at 4 °C.

4. Retrieval of polyacrylamide beads. Centrifuge for 5 min at 3000 r.p.m. in a refrigerated centrifuge and keep the supernatant for analysis. Gently resuspend the pellet in 1 ml PBS⁻/BSA to prevent vesicularization, and repeat the centrifugation step. (For analysis by gel electrophoresis, the last wash is in PBS⁻ without BSA.)

5. Retrieval of magnetic beads. Use a small permanent magnet placed at the side of the lower portion of the tube on ice to collect the beads (keep the supernatant for analysis). Gently resuspend in 1 ml PBS⁻/BSA to prevent vesicularization and repeat the same step.

6. Resuspend the immunoadsorbent plus bound vesicles in 100 μl of the same buffer as used in the assay (for example, homogenization buffer) and proceed immediately with the cell-free assay (*Protocol 11*).

4.5 Preparation of the cytosol

To prepare cytosol, it is convenient to use the same homogenization conditions as described in *Protocol 5* (18, 22). With 10–20 dishes, a PNS is obtained in a volume of ≈ 2 ml in 250 mM sucrose, 3 mM imidazole, pH 7.4. The PNS is centrifuged at 22 000 *g* for 20 min and the supernatant is recentrifuged at 185 000 *g* for 60 min. The high-speed supernatant, which corresponds to the cytosol, is expected to contain 10–15 mg/ml, a range appropriate for the fusion assay. The cytosol is aliquoted (50 μl), frozen and stored in liquid nitrogen. If the homogenization buffer is not appropriate for some experiment, it can be exchanged using a G25 column with a buffer often used in the fusion assay (e.g. 85 mM sucrose; 1.5 mM $Mg[OAc]_2$; 1 mM dithiothreitol (DTT); 100 mM KOAc; 10 mM Hepes, pH 7.4). However, this procedure will also deplete small cytosolic components.

Alternatively, cytosol can be prepared directly in this buffer but homogenization may be more difficult, requiring the use of a tight-fitting Dounce homogenizer. Cytosols prepared with these protocols support fusion with a similar efficiency *in vitro*; thus, the choice depends largely on the type of experiments.

5. Cell-free assay of endocytic vesicle fusion

5.1 Components

The different cellular components of the assay are the avidin- and the bHRP-labelled fractions, as well as the cytosol (see outline in *Figure 1*). Different combinations of fractions can be used: PNS (*Protocol 5*), fractions obtained from the gradient (*Protocol 9*), or immunoisolated fractions (*Protocol 10*). In typical experiments, we use 300 μg protein from each PNS or 15–25 μg protein from each fraction recovered from the flotation gradient. In the assay, cytosol is generally present at concentrations varying from 1 to 10 mg/ml. The efficiency of fusion is, however, higher at a high cytosol concentration, presumably because cytosolic factors required for the fusion process are then more abundant (22).

5.2 Experimental conditions and requirements

The cellular components of the fusion reaction are added sequentially to a microcentrifuge tube on ice, which already contains salts, buffer, ATP, and biotinylated insulin (to quench avidin that may be free in the mixture). The reagents needed are listed and the experiment is outlined in *Protocol 11*. The fusion process is strictly dependent on ATP. Moreover, K^+, most commonly used as a KCl or KOAc salt, must be present within a concentration range ≈ 50–100 mM. Hypertonic conditions, however, have little influence on the fusion process, up to ≈ 500–600 mOsm. The tube is then left for 2–3 min on ice in order to reduce the effects of osmotic and ionic changes occurring in the medium (14). The tube containing the reaction mixture is then warmed to the desired temperature. Our group and others have observed that little if any fusion occurs at temperatures below 20 °C and that fusion activity increases with increasing temperatures up to 37 °C (see reference 2). When increasing the length of the reaction at 37 °C, the extent of fusion increases linearly for the first 20–30 min, depending on the concentration of the cellular components, and then a plateau is slowly reached. At the end of the incubation, the tube is cooled to ice temperature, extracted in detergent after adding excess biotinylated insulin, and the avidin–bHRP complex that has formed upon fusion of endocytic vesicles is immunoprecipitated.

5.3 Measurement of fusion efficiency

The total amount of avidin present in the fraction is quantified using the same detection system (*Protocol 11*). Biotin–insulin is then omitted in the assay and replaced by saturating amounts of bHRP during detergent solubilization. The amount of avidin complexed to bHRP during fusion is then expressed as a percentage of the total amount of avidin which can be complexed to bHRP. This calculation measures the efficiency of the fusion process, provided that avidin–bHRP is quantitatively immunoprecipitated. With fluid phase markers, this efficiency reflects the extent of mixing between the content of labelled endosomal vesicles in the assay, an estimate of the fusion activity of the vesicles. It should be noted that internalized avidin coupled to cell surface biotinylated proteins may exhibit a different distribution in endosomes compared to that of free avidin, particularly at later stages of the pathway. The total amount of avidin in the fraction and the fusion efficiency should always be determined, as well as the total bHRP activity added to the assay, to allow comparison between different experiments (amount of fractions, time, temperature, cytosol concentration, specific reagents, etc.).

As an example, *Figure 4* shows the compared fusion efficiencies of early endosomes prepared either from the non-polarized BHK cells or polarized epithelial MDCK cells. The fractions were obtained from the flotation gradient (*Protocol 9*). The assay was carried out either in the presence of BHK or MDCK cytosols at the same protein concentration (≈ 3 mg/ml in the assay, prepared as described in Section 4.5) or in the presence of *Xenopus* cytosol at a high protein concentration (≈ 10 mg/ml in the assay; see reference 19). Fusion efficiency (see *Protocol 11*) is in a similar range with BHK or MDCK cytosol, but increased by the high protein concentration of *Xenopus* cytosol. More important, BHK early endosomes fuse only with basolateral and not with apical MDCK early endosomes. (In this latter experiment, MDCK cells were grown on plastic dishes. Basolateral early endosomes were labelled after opening the tight junctions in PBS without divalent cations (PBS^-; *Protocol 3*); in the presence of divalent cations only apical endosomes were labelled.) These observations agree well with our findings that apical and basolateral early endosomes in MDCK cells use different mechanisms of recognition/fusion (18), and suggest that early endosomes from the non-polarized BHK cells can only interact with basolateral early endosomes.

Protocol 11. Cell-free assay of endocytic vesicle fusion

Materials

- biotinylated insulin (see *Protocol 4*)
- 0.625 M Hepes (titrated to pH 7.0 with KOH); 75 mM $Mg(OAc)_2$; 50 mM DTT. Aliquoted and stored at −20 °C (50 × concentrated stock)
- 1.0 M KOAc aliquoted and stored at −20 °C (20 × concentrated stock)
- sterile stock solution of bHRP (e.g. 1.8 mg/ml) stored at 4 °C

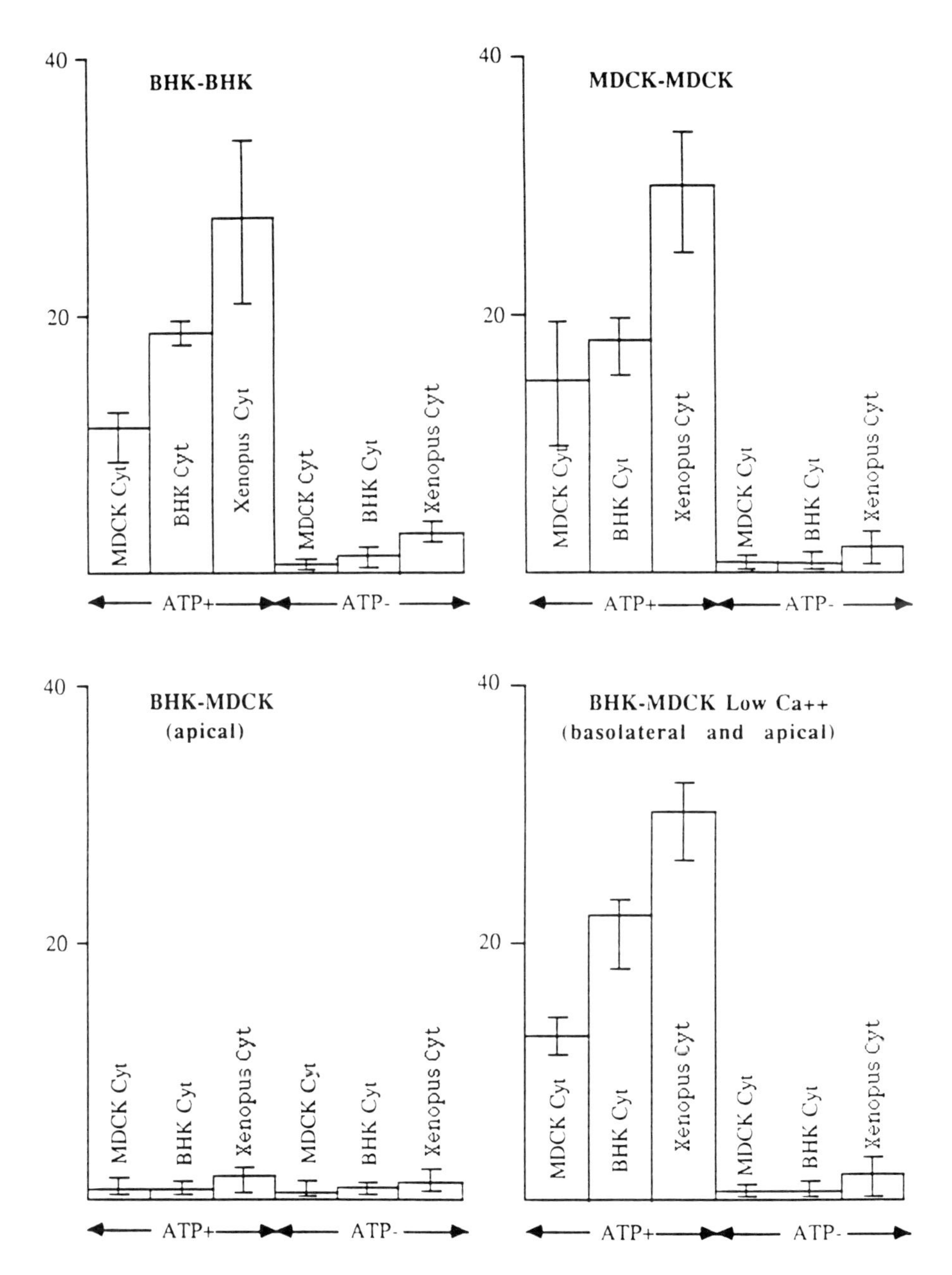

Figure 4. Fusion assay with BHK and MDCK early endosomal fractions in different cytosols.

Protocol 11. *Continued*

- 100 mM ATP titrated to pH 7.0 with NaOH. Stored in 20- or 50-μl aliquots at −20 °C
- 800 mM creatine phosphate. Stored in 20- or 50-μl aliquots at −20 °C
- 4 mg/ml creatine phosphokinase (800 U/ml, Boehringer) in 50% glycerol. Stored in 20- or 50-μl aliquots at −20 °C
- $(NH_4)_2SO_4$ precipitate of hexokinase stored at 4 °C (1400 U/ml, Boehringer)
- 0.25 M D-glucose
- 10% TritonX-100 at 4 °C
- 0.17 mg/ml bHRP (bHRP stock solution diluted with PBS^-)
- PBS^-/BSA (*Protocol 2*)
- reagents and materials for avidin immunoprecipitation and bHRP detection (*Protocols 7* and *8*)
- water bath at 37 °C

Method

1. Prepare the ATP regenerating system at 4 °C just before the experiment by mixing 1:1:1 volumes of the following stock solutions: 100 mM ATP; 800 mM creatine phosphate; 4 mg/ml creatine phosphokinase
2. Prepare the ATP depleting system at 4 °C just before the experiment from $(NH_4)_2SO_4$ precipitate of hexokinase stored at 4 °C. Vortex the suspension, pipette the desired amount (e.g. 50 μl ≈ 0.5 mg), centrifuge for 5 min in a microcentrifuge, and aspirate the supernatant. The pellet is dissolved in the same volume, i.e. 50 μl, of 0.25 M D-glucose.
3. Mix in a 1.5-ml microcentrifuge tube on ice:
   - 3.0 μl of Hepes/$Mg(OAc)_2$/DTT (50 × stock)
   - 7.5 μl of 1 M KOAc (20 × stock)
   - 8.0 μl biotinylated insulin (*or* omit biotinylated insulin for the detection of the total avidin in the fraction)
   - plus ATP: 10 μl ATP regenerating system (step **1**) (*or* minus ATP: 8 μl ATP depleting system (step **2**))
4. Then add sequentially in each tube and mix gently: 50 μl cytosol + 50 μl avidin-labelled fraction + 50 μl bHRP-labelled fraction. Leave for 2–3 min on ice.
5. Incubate the tube at 37 °C (e.g. 45 min). Do not agitate or mix.
6. Cool to ice temperature.
7. When using PNS or fractions from the gradient:

(a) Add 5 µl biotinylated insulin (*or* 10 µl of 0.17 mg/ml bHRP for the detection of the total avidin in the fraction) and mix well.

(b) Add 7.5 µl Triton X-100.

(c) Vortex, leave on ice for 30 min, and then dilute to 1 ml with PBS^-/BSA.

When using immunoisolated fractions:

(a) Add 1 ml PBS^-/BSA and centrifuge for 5 min at 3000 r.p.m. (polyacrylamide beads) *or* retrieve with a magnet (magnetic beads).

(b) Resuspend the beads in 50 µl PBS^-/BSA, add 5 µl biotinylated insulin (*or* 10 µl of 0.17 mg/ml bHRP for the detection of the total avidin in the fraction) and mix well.

(c) Add 5 µl Tritin X-100.

(d) Vortex, leave on ice for 30 min, and then dilute to 1 ml with PBS^-/BSA.

8. Centrifuge at 5000 r.p.m. for 3 min. Collect supernatant.

9. Quantify avidin–bHRP in the supernatant (*Protocols 8* and *9*).

---

6. Conclusion

When studying intracellular events, *in vitro* approaches have the major advantage that experimental conditions can be manipulated in the assay. These manipulations can, in fact, be carried out separately on the subcellular fractions and the cytosol before, during, or after the assay. The cytosol dependence of these assays in particular has been instrumental in demonstrating the involvement of a factor which is also required for transport in the Golgi (24, 26), the effect of mitotic cytosol and of the cdc2 kinase (19), and the role of *rab5* in early endosome fusion (22). The assay has also been used to study antigens involved in the recognition/fusion process, either by antibody-mediated inactivation of a component present on the cytoplasmic face of the membrane (22) or by immunodepletion of a cytosolic protein (18, 19, 22, 26). The long-standing interest in cell-free assays comes largely from the possibility they offer of reducing the complex series of events involved in the recognition/fusion process to a series of biochemical reactions that can be dissected and studied in the test-tube.

References

1. Helenius, A., Mellman, I., Wall, D., and Hubbard, A. (1983). *Trends biochem. Sci.* **8**, 245.
2. Gruenberg, J. and Howell, K. E. (1989). *Ann. Rev. Cell Biol.* **5**, 453.
3. Kornfeld, S. and Mellman, I. (1989). *Ann. Rev. Cell Biol.* **5**, 483.

4. Gruenberg, J., Griffiths, G., and Howell, K. E. (1989). *J. Cell Biol.* **108**, 1301.
5. Dunn, W. A., Connolly, T. P., and Hubbard, A. L. (1986). *J. Cell Biol.* **102**, 24.
6. Rothman, J. E. and Orci, L. (1990). *FASEB J.* **4**, 1460.
7. Smythe, E., Pypaert, M., Lucocq, J., and Warren, G. (1989). *J. Cell Biol.* **108**, 843.
8. Podbilewicz, B. and Mellman, I. (1990). *EMBO J.* **9**, 3477.
9. Goda, Y. and Pfeffer, S. R. (1988). *Cell* **55**, 309.
10. Mullock, B. M., Branch, W. J., van Schaik, M., Gilbert, L. K., and Luzio, J. P. (1989). *J. Cell Biol.* **108**, 2093.
11. Davey, J., Hurtley, S. M., and Warren, G. (1985). *Cell* **43**, 643.
12. Gruenberg, J. and Howell, K. E. (1986). *EMBO J.* **5**, 3091.
13. Gruenberg, J. and Howell, K. E. (1987). *Proc. nat. Acad. Sci., USA* **84**, 5758.
14. Braell, W. A. (1987). *Proc nat. Acad. Sci., USA* **84**, 1137.
15. Diaz, R., Mayorga, L., and Stahl, P. (1988). *J. Biol. Chem.* **263**, 6093.
16. Woodman, P. G. and Warren, G. (1988). *Eur. J. Biochem.* **173**, 101.
17. Beaumelle, B. D. and Hopkins, C. R. (1989). *Biochem. J.* **264**, 137.
18. Bomsel, M., Parton, R., Kuznetsov, S. A., Schroer, T. A., and Gruenberg, J. (1990). *Cell* **62**, 719.
19. Tuomikoski, T., Felix, M.-A., Dorée, M., and Gruenberg, J. (1989). *Nature* **342**, 942.
20. Warren, G. (1985). *Trends biochem. Sci.* **502**, 439.
21. Mayorga, L. S., Diaz, R., and Stahl, P. D. (1989). *Science* **244**, 1475.
22. Gorvel, J.-P., Chavrier, P., Zerial, M., and Gruenberg, J. (1991). *Cell* **64**, 915.
23. Chavrier, P., Parton, R. G., Hauri, H. P., Simons, K., and Zerial, M. (1990). *Cell* **62**, 317.
24. Block, M. R., Glick, B. S., Wilcox, C. A., Wieland, F. T., and Rothman, J. (1988). *Proc. nat. Acad. Sci., USA* **85**, 7852.
25. Beckers, C. J. M., Block, M. R., Glick, B. S., Rothman, J. E., and Balch, W. E. (1989). *Nature* **339**, 397.
26. Diaz, R., Mayorga, L., Weidman, P. J., Rothman, J. E., and Stahl, P. D. (1989). *Nature* **339**, 398.
27. Gelfand, V. I. (1989). *Curr. Opinions Cell Biol.* **1**, 63.
28. Howell, K. E., Schmid, R., Ugelstad, J., and Gruenberg, J. (1989). *Methods Cell Biol.* **31A**, 264.
29. Griffiths, G., Hoflack, B., Simons, K., Mellman, I., and Kornfeld, S. (1988). *Cell* **52**, 329.
30. Howell, K. E., Devaney, E., and Gruenberg, J. (1989). *Trends biochem. Sci.* **14**, 44.
31. Orr, G. A., Heney, G. C., and Zeheb, R. (1986). In *Methods in Enzymology*, Vol. 122, (ed. F. Chytil and D. B. McCormick), p. 83. Academic Press, London.
32. Davoust, J., Gruenberg, J., and Howell, K. E. (1987). *EMBO J.* **6**, 3601.
33. Samuelson, A. C., Stockert, R. J., Novikoff, A. B., Novikoff, P. M., Saez, J. C., Spray, D. C., and Wolkoff, A. W. (1988). *Am. J. Physiol. 254*, 829.
34. Sandvig, K., Olsnes, S., Petersen, O. W., and van Deurs, B. (1987). *J. Cell Biol.* **105**, 6799.
35. Cosson, P., de Curtis, I., Pouysségur, J., Griffiths, G., and Davoust, J. (1989). *J. Cell Biol.* **108**, 377.
36. Heuser, J. (1989). *J. Cell Biol.* **108**, 855.

37. Parton, R. G., Dotti, C. G., Bacallao, R., Kurtz, I., Simons, K., and Prydz, K. (1991). *J. Cell Biol.* **113**, 261.
38. Le Bivic, A., Real, F. X., and Rodriguez-Boulan, E. (1989). *Proc. nat. Acad. Sci., USA* **86**, 9313.
39. Bretscher, M. S. and Lutter, R. (1988). *EMBO J.* **7**, 4087.
40. Beaufay, H. and Amar-Costesec, A. (1976). In *Methods in membrane biology*, Vol, 6 (ed. E. D. Korn), pp. 1–99. Plenum Press.
41. Morré, D. J., Howell, K. E., Cook, G. M. W., and Evans, W. H. (ed.) (1988). *Prog. clin. Biol. Res.* **270**.
42. Luzio, J. P., Mullock, B. M., Branch, W. J., and Richardson, P. J. (1988). *Prog. clin Biol. Res.* **270**, 77.
43. Kreis, T. E. (1986). *EMBO J.* **5**, 931.

9

In vitro reconstitution of vesicular transport from the endoplasmic reticulum to the *cis* Golgi in semi-intact cells

RUTH SCHWANINGER

1. Introduction

Proteins destined for the plasma membrane, lysosomes, and endosomes, or the extracellular fluid are transported between organelles of the exocytic or secretory pathway of eukaryotic cells by vesicles that bud from one compartment and fuse with another. A high degree of specificity for this sequential budding, targeting, and fusion of transport vesicles must be maintained, or the cell would rapidly destroy its highly compartmental organization.

Progression of newly synthesized proteins along this pathway can be followed by observing the maturation of N-linked oligosaccharide chains of glycoproteins during their passage through the different organelles. This is possible as individual oligosaccharide-processing enzymes reside in particular subcellular compartments.

In order to identify and characterize components of the molecular machinery involved in protein transport between successive compartments, its reconstitution in cell-free preparations is required. Typically this involves the transfer of a marker protein from a donor to an acceptor compartment and is followed by a specific modification confined to the target compartment. Several systems that reconstitute protein transport in the endocytic pathway (reviewed in references 1 and 2) and at various stages of the exocytic pathway (reviewed in references 2–4) have been developed, individual donor and acceptor membranes being provided by cell homogenates (5–10) or partially purified subcellular fractions (11, 12). In alternative cell-free assays, the donor and the acceptor compartment exist in the context of a semi-intact cell, which ensures minimal disruption of cellular organization (13–18). In each case, the transport reaction depends upon the presence of cytosol (a high-speed cell supernatant), ATP, and incubation at temperatures in excess of

15 °C. This chapter will focus on transport of protein between the endoplasmic reticulum (ER) and the Golgi in semi-intact cells.

2. Protein transport in semi-intact cells

Semi-intact (or perforated) cells are a population of cells that have lost portions of their plasma membrane as a result of physical perforation. Intracellular organelles such as the nucleus, ER, and Golgi are preserved in an intact and functional form within the cell. These cells lose their cytoplasmic contents, so that the composition of the cytosol can be manipulated by fractionation and reconstitution of its component parts. Furthermore, the secretory organelles are accessible to exogenous factors including inhibitors, enzymes, and antibodies, allowing the identification of components involved in intracellular transport.

The system described in Section 3 measures ER to *cis* Golgi transport (13, 19, 20), and can also be modified to allow measurement of transport from ER to *cis* Golgi and successively to *medial* Golgi (18, 19). In principle, transport of any secretory or plasma membrane protein can be followed. We follow the transport of vesicular stomatitis virus (VSV) G-protein, which is transported to the cell surface of virus-infected cells through the same pathway as endogenous plasma membrane proteins. The virus infection inhibits synthesis of cellular proteins, so that it is possible to measure labelled VSV G-protein directly by loading the complete cell lysate on a gel, without the need to immunoprecipitate the protein of interest. Transport is measured by following the maturation of the two oligosaccharide chains on VSV G-protein during transit through the various secretory compartments (*Figure 1*).

2.1 Processing of glycoproteins in the early exocytic pathway

The *cis* Golgi localized α-1,2-mannosidase I (Man I) trims the $man_9glcNAc_2$ oligosaccharide (the ER form) to the $man_5glcNAc_2$ form, which is uniquely sensitive to cleavage of the oligosaccharide chains by the enzyme endo-glycosidase D (endo D) (*Figure 1*). The endo D sensitive *cis* Golgi form (G_D) can be distinguished from the endo D resistant ER form (G_{D0}) by post-incubation of lysed semi-intact cells with endo D and separation of VSV-G protein by SDS–PAGE (*Figure 2*). The $man_5glcNAc_2$, form is further processed to forms that are again endo D resistant.

3. ER to *cis* Golgi transport

Our *in vitro* system has been developed with factors that simplify the analysis of transport: Chinese hamster ovary (CHO) cells are infected with VSV. For

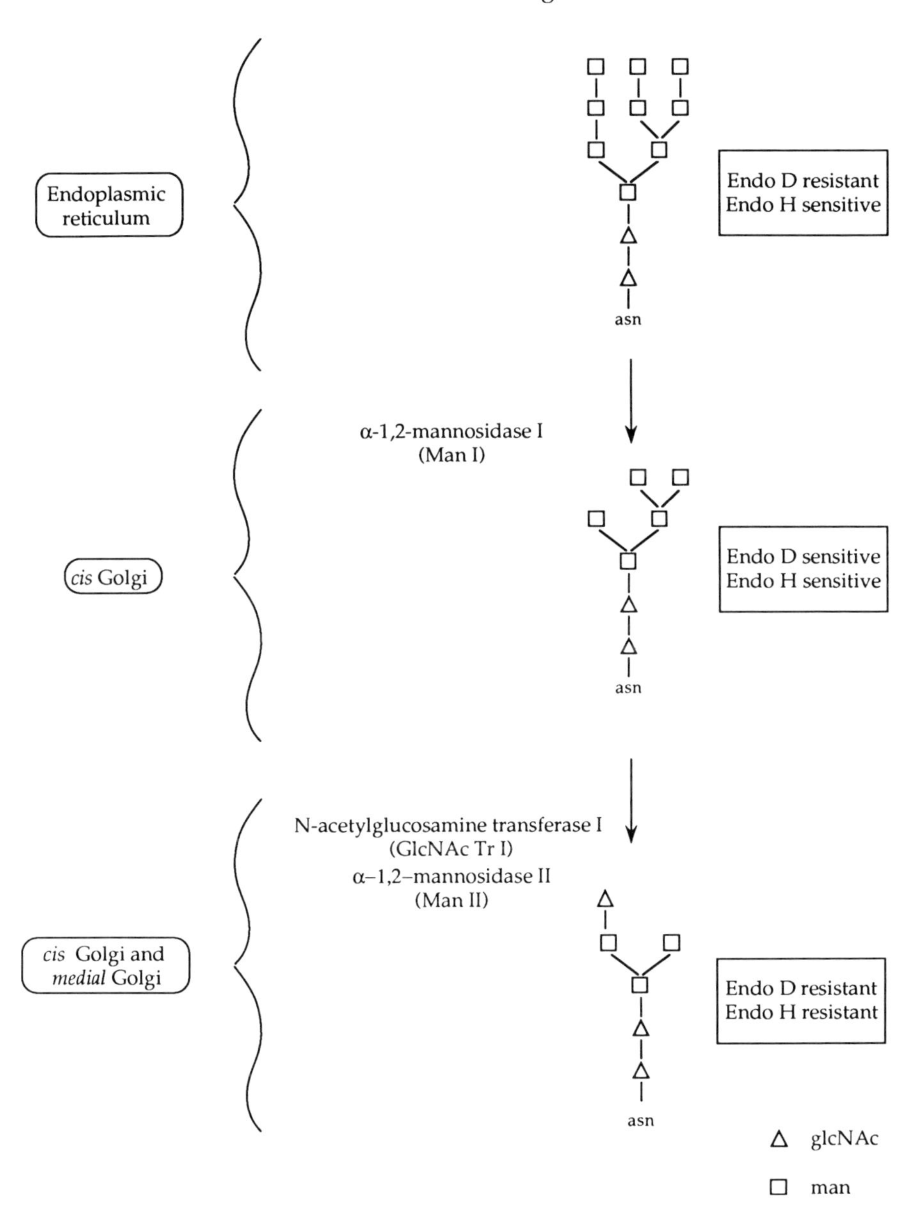

Figure 1. Glycosylation steps in the early exocytic pathway.

this assay we routinely use VSV ts045, a temperature-sensitive VSV mutant. The ts045 G-protein is retained within the ER at 40 °C, the restrictive temperature. This transport inhibition is reversible at 32 °C, the permissive temperature. To ensure that all VSV G-protein is localized in the ER at the

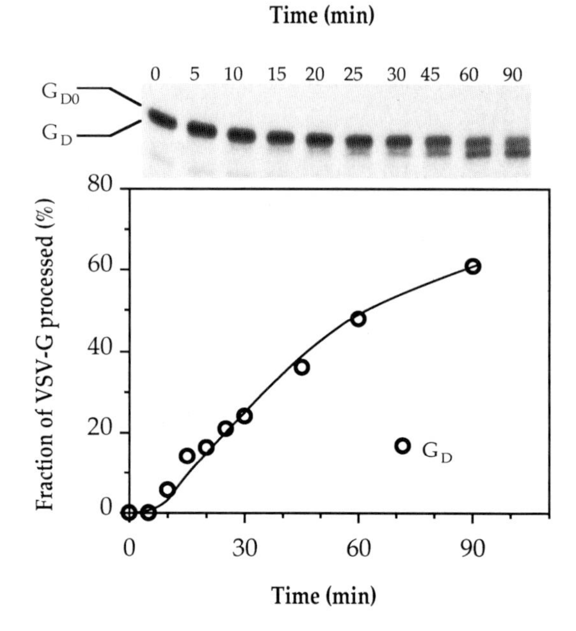

Figure 2. Time course of ER to *cis* Golgi transport measured by acquisition of endo D sensitivity.

beginning of the *in vitro* transport incubation, ts045-infected CHO cells are radioactively labelled at the restrictive temperature. This assay can also be done with wild-type VSV. In this case, the labelling time is shorter (2–4 min), the amount of [^{35}S]methionine is doubled, and the chase with cold methionine is omitted.

The endo D sensitive protein is only a transient intermediate in the exocytic pathway, and is immediately further processed to forms that are again endo D resistant (*Figure 1*). To be able to measure all G protein that has been processed by Man I, we use the 15B mutant cell line of CHO. These cells are defective in *N*acetylglucosamine transferase I (glcNAc TrI), the next enzyme in the glycosylation pathway, so that all protein that has reached or progressed beyond the *cis* Golgi compartment remains endo D sensitive.

The assay can also be done in wild-type cells (see Section 2.2.4). As we gel filter the cytosol to remove low molecular weight compounds, in the absence of UDP-glcNAc, the substrate for the next oligosaccharide-processing enzyme in the pathway, only a small fraction of VSV G-protein is further processed to forms that are again endo D sensitive.

3.1 Preparation of VSV-infected, labelled semi-intact cells

The CHO 15B cells are maintained in monolayer culture on 10-cm dishes in α-MEM (α-minimal essential medium: Earle's salts, with glutamine and nucleosides), supplemented with 100 IU/ml penicillin, 100 μg/ml streptomycin, and 8% fetal calf serum (FCS). The cells are passaged such that at the time of infection they form a complete monolayer while still maintaining a well-spread morphology.

On the morning of use cells are infected with virus at the permissive temperature (see *Protocol 2*), and then viral proteins are labelled *in vivo* at the restrictive temperature (see *Protocol 3*). Virus stocks are propagated and stored as described in *Protocol 1*.

Protocol 1. Propagation of virus

Materials

- baby hamster kidney (BHK)[a] cells, grown in monolayer on 10-cm dishes in G-MEM (minimal essential medium), 5% FCS, 10% tryptose phosphate broth (TPB)
- VSV
- G-MEM (minimal essential medium, Glasgow), 10% TPB
- TD buffer: 138 mM NaCl, 5 mM KCl, 25 mM Tris base, 0.4 mM Na_2HPO_4; pH 7.2

Method

1. Wash BHK cells with TD.
2. Add per 10 cm dish: 10 ml G-MEM; 10% TPB, containing approximately 10^6 plaque-forming units (PFU), of VSV (multiplicity of infection = 0.1).
3. Incubate 48 h (until cells begin to round up) at 32 °C (for ts045) or at 37 °C (for wild-type VSV).
4. Remove virus containing supernatant.
5. Centrifuge supernatant at 800 *g* for 10 min to remove cells.
6. Freeze cell-free supernatant in 100-μl aliquots in liquid nitrogen.
7. Store at −80 °C.

[a] For infection of some cell lines, it is necessary for the virus to be adapted to that cell line to achieve a good infection. In this case, use the cell line that is to be infected instead of BHK cells for the virus propagation described above.

Protocol 2. Infection of CHO 15B cells with VSV

Materials

- rocking platform in 32 °C CO_2-incubator
- infection-medium: α-MEM; 25 mM Hepes–KOH, pH 7.2

Protocol 2. *Continued*

- 1 mg/ml actinomycin D in ethanol (store at −20 °C)
- VSV stock solution (approximately 2 × 10^9 p.f.u./ml, see *Protocol 1*)
- post-infection medium: α-MEM; 8% FCS

Method

1. Thaw virus at 32 °C.
2. Mix infection-cocktail:
   - 1 ml infection-medium, prewarmed to 32 °C
   - 5 μl actinomycin D
   - 100 μl of virus stock solution
3. Remove the medium of a 10-cm tissue culture dish with cells that have just reached confluency.
4. Add the infection cocktail to this dish; move to distribute evenly.
5. Incubate 45 min at 32 °C with constant rocking of the dish.
6. Add 5 ml post-infection medium to the dish.
7. Incubate 3.5–4.5 h at 32 °C (for ts045-infected cells) or at 37 °C (for wild-type VSV-infected cells).

Protocol 3. Radioactive labelling of VSV-infected CHO cells

Materials

- 40 °C water bath with a perforated stainless steel platform situated just below the surface of the water. The depth of the water (a few mm above the platform) is just sufficient to immerse the base of the tissue culture dish without the dish floating in the water when the lid is removed
- labelling medium (methionine-deficient; Sigma, M 7270), supplemented with leucine, lysine, and 25 mM Hepes–KOH, pH 7.2, pre-warmed to 40 °C
- [^{35}S]methionine (Translabel, ICN Biomedicals, 51006)
- 0.25 M methionine

Method

Note: It is essential that the temperature of the cells does not drop below 40 °C during the labelling procedure to ensure that labelled ts045 G-protein remains in the ER.

1. Place the dish in the 40 °C waterbath and aspirate the medium.

2. Wash the cells twice with 3-ml portions of 40 °C pre-warmed labelling medium.
3. Incubate 5 min with 5 ml of 40 °C labelling medium.
4. Replace the medium with another 1.5 ml labelling medium and add 10 μl (100 μCi) of [^{35}S]methionine; rock the plate gently to ensure even coverage of the cells.
5. Incubate 10 min at 40 °C, rocking the dish briefly at about 2-min intervals to prevent the cells from drying out. Maintain contact with the waterbath while rocking.
6. Add 30 μl of 0.25 M methionine and incubate for 2 min at 40 °C.

To perforate the infected labelled cells we routinely use the hypotonic swelling method (*Protocol 4*). This results in more than 95% semi-intact cells in the case of CHO 15B cells, and about 90% semi-intact cells in the case of wild-type CHO cells. The swelling procedure can be adjusted to different cell types by changing the swelling time, or by washing the cells in the swelling buffer instead of the assay buffer before swelling. Semi-intact cells are permeable to trypan blue, a membrane-impermeant chromatin-binding dye, and are in this way readily distinguishable from intact cells.

In some cases it may not be preferable to subject the cells to hypotonic buffer, and also some cell lines are not perforated efficiently with the hypotonic swelling method. An alternative method is to attach the cells more firmly to the culture dish by pre-treating it with poly-L-lysine, such that non-swollen cells also become semi-intact by scraping (*Protocol 5*).

An alternative method for preparing semi-intact cells using nitrocellulose stripping is described in references 13 and 14. Growing cells are covered with a sheet of nitrocellulose paper, the sheet dried, and then peeled off, tearing pieces of plasma membrane from the cells.

Protocol 4. Preparation of semi-intact cells by hypotonic swelling

Materials

- ice-water bath with a stainless steel platform situated just below the surface of the water (see *Protocol 3*)
- rubber policeman
- assay buffer: 50 mM Hepes–KOH, pH 7.2; 90 mM potassium acetate (KOAc)
- swelling buffer: 10 mM Hepes–KOH, pH 7.2; 18 mM KOAc
- 1% trypan blue in water

Protocol 4. *Continued*

Method

1. Aspirate the medium from the labelled cells, transfer the dish immediately to the ice-water bath, and add 5 ml cold assay buffer to cool the cells rapidly and to prevent transport.
2. Wash the cells two times with cold assay buffer.
3. Overlay the cells with 5 ml swelling buffer.
4. Let stand 10 min on ice.
5. Aspirate swelling buffer.
6. Add 3 ml cold assay buffer.
7. Scrape the cells immediately with the rubber policeman, using firm strokes.
8. Transfer the cells with a Pasteur pipette into a 15-ml polypropylene centrifuge tube.
9. Pellet for 3 min at 800 *g* at 4 °C.
10. Wash the pellet with 3 ml assay buffer.
11. Pellet for 3 min at 800 *g* at 4 °C.
12. Resuspend the pellet in 200–300 μl assay buffer.
13. Check index of semi-intact cells: mix 10 μl cell suspension and 1 μl 1% trypan blue on coverslip; count cells with blue-stained nucleus using phase contrast microscopy.

Protocol 5. Preparation of semi-intact cells on poly-L-lysine coated plates

Materials

- rubber policeman
- 5 mg/ml poly-L-lysine stock solution, filter-sterilized, stored at 4 °C
- 5 μg poly-L-lysine/ml H_2O
- TD buffer (see *Protocol 1*)
- assay buffer (see *Protocol 4*)

Method

1. Wash culture dish with 10 ml sterile H_2O.
2. Add 2.5 ml poly-L-lysine (5 μg/ml) to culture dish.
3. Incubate 60 min at 32 °C.
4. Wash dish three times with TD buffer.

5. Plate cells in regular medium on pre-treated dish.
6. Incubate for 1–2 days until cells reach confluency.
7. Wash cells in assay buffer.
8. Aspirate buffer.
9. Scrape cells with rubber policeman.
10. Harvest cells into 10 ml assay buffer.
11. Pellet 3 min at 800 *g* at 4 °C.
12. Resuspend cells in 200–300 μl assay buffer.

3.2 *In vitro* transport reaction to the endo D sensitive form

The semi-intact cells are prepared fresh. They can be stored on ice after preparation for up to 3 hours, although the transport efficiency is best if the cells are incubated within 1 h. Transport can be reconstituted by incubation in the presence of cytosol (a 100 000 *g* supernatant of CHO cell homogenate, see *Protocol 6*) and an ATP-regenerating system (*Protocol 7*).

Protocol 6. Preparation of a cytosol fraction (100 000 *g* supernatant)

Materials

- ball-bearing homogenizer (11, 21)
- rubber policeman
- 40 15-cm dishes with confluent CHO cells (wild-type or 15B)
- TEA buffer: 125 mM KOAc; 10 mM triethanolamine (TEA)-HCl, pH 7.2
- 25/125 Hepes/KOAc: 25 mM Hepes–KOH, pH 7.2; 125 mM KOAc
- Sephadex G-25 column

Method

1. Aspirate most of the medium from the CHO cell cultures; scrape cells gently with a rubber policeman in remaining medium.
2. Pellet the cells 5 min at 800 *g* at 4 °C.
3. Resuspend the cells in 15 ml TEA buffer.
4. Pellet the cells 5 min at 800 *g* at 4 °C.
5. Determine the volume of the cell pellet and add 4 volumes of 25/125 Hepes/KOAc.
6. Homogenize the cells with the ball-bearing homogenizer.
7. Centrifuge for 60 min at 100 000 *g*.

Protocol 6. *Continued*

8. Remove the whitish lipid layer at the top of the tube carefully by aspiration.
9. Remove the supernatant (cytosol), taking care not to disturb the pellet.
10. Desalt the cytosol: load on a Sephadex G-25 column in 25/125 Hepes/KOAc.
11. The majority of the cytosolic proteins will come off the column in the void volume.
12. Determine the protein concentration of the fractions.
13. Pool the peak fractions.
14. Concentrate if the concentration is less than 5 mg/ml.
15. Freeze the cytosol in about 100-µl aliquots in liquid nitrogen and store at −70 °C.

Protocol 7. *In vitro* **reconstitution of ER to** ***cis*****-Golgi transport in semi-intact cells**

Materials

- semi-intact cells in 50 mM Hepes, 90 mM KOAc; these must be fresh (not older than 3 h; see *Protocol 4*)
- cytosol in 25 mM Hepes, 125 mM KOAc; stored at −70 °C (see *Protocol 6*)
- ATP (Na form): 20 mM
- creatine phosphate (CP): 100 mM
- creatine phosphokinase (CPK): 0.2 IU (CPK)/2 µl
- 100 mM magnesium acetate (MgOAc), pH 7
- 1 M KOAc
- $(NH_4)_2SO_4$ precipitate of hexokinase stored at 4 °C (1400 U/ml, Boehringer)
- 1 M Hepes–KOH, pH 7.2
- 10× Ca^{2+}/EGTA buffer: 50 mM EGTA; 18 mM $CaCl_2$; 20 mM Hepes–KOH, pH 7.2

Method

1. Prepare a 20 × stock solution of ATP-regenerating system in the following way.
   (a) Mix ATP (Na form) and CP in approximately 16 ml H_2O. Adjust pH to 7.0.

(b) Dissolve CPK in 2 ml H_2O and add to ATP/CP. Adjust volume to 20 ml and store in 100-μl portions at −70 °C.

2. Mix the transport cocktail in 1.5-ml microcentrifuge tubes[a] on ice. Add ingredients from bottom to top of the following list, beginning with water (to make a final volume of 40 μl) and ending with the semi-intact cells. When inhibitors or other factors like antibodies or purified proteins are added, unspecific effects of the buffer in which the factor is added have to be controlled by adding the buffer alone to a control tube.

Solution	*Volume μl*	*Final concentration*
semi-intact cells	5	ca.25 μg protein; 11 mM KOAc; 6 mM Hepes
cytosol	5	ca.25 μg protein; 16 mM KOAc; 3 mM Hepes
ATP-regenerating system	2	1 mM ATP; 5 mM CP; 0.2 IU CPK
100 mM MgOAc	1	2.5 mM MgOAc
1 M KOAc	2	50 mM KOAc
1 M Hepes–KOH, pH 7.2	1	25 mM Hepes
10 × Ca^{2+}/EGTA buffer	4	5 mM EGTA; 1.8 mM Ca^{2+} (100 nM free Ca^{2+}); 2 mM Hepes
H_2O	to 40 μl	

The total transport cocktail has a volume of 40 μl with a final concentration of 77 mM KOAc and 36 mM Hepes, pH 7.2. Vortex the transport cocktail gently for about 1 sec before incubation.

3. Incubate 90 min at 32 °C (for VSV ts045 infected cells) or at 37 °C (for wild-type VSV infected cells).

4. Transfer the tubes to ice to stop the transport reaction.

[a] Sarstedt, cat. no. 72.690. Some brands may contain chemical residues or surfactants which inhibit transport and should be compared to Sarstedt tubes before use.

3.3 Detection and quantitation of transport

To make it easier to distinguish between the $man_5glcNAc_2$ *cis* Golgi-form and the $man_9glcNAc_2$ ER-form, the cells are pelleted, lysed, and digested with endo D. This enzyme cleaves the $man_5glcNAc_2$ oligosaccharide, leaving one glcNAc molecule per oligosaccharide chain bound to the protein. The endo D digested glycoprotein (G_D, *Figure 2*) has a faster mobility on SDS–PAGE than the undigested $man_9glcNAc_2$ form (G_{D0}, *Figure 2*), making it possible to assess the amount of protein that has been transported.

For digestion of glycoprotein in 15B cells, we prepare endo D from *Diplococcus pneumoniae* as described in reference 22. When using wild-type cells, pure endo D should be used to prevent unspecific digestion of higher glycosylated forms by impurities present in our endo D preparations.

Protocol 8. Post-incubation with endo D

Material

- endo D buffer: 50 mM phosphate buffer, pH 6.5 (prepared by mixing NaH_2PO_4 and Na_2HPO_4 until pH 6.5 is reached); 5 mM EDTA; 0.2% Triton- X-100
- endo D: endo D from *D. pneumonia* (22) for CHO-15B; pure endo D (Boehringer Mannheim) for wild-type cells

Method

1. Pellet the cooled cells by centrifugation for 15 sec in a microcentrifuge at 15 000 *g*.
2. Resuspend the pellet in 40 μl endo D buffer and 3 μl endo D.
3. Incubate overnight at 37 °C.

The fraction of VSV G-protein that has been transported to the *cis*-Golgi is determined using sodium dodecyl sulphate (SDS)-polyacrylamide gel electrophoresis (SDS–PAGE), fluorography, and densitometry.

Protocol 9. Sample analysis: SDS–PAGE, fluorography, and densitometry

Materials

- 5 × concentrated gel sample buffer (5 × GSB): 31.25 ml 1 M Tris, pH 6.8; 50 ml glycerol; 1.25 g dithiothreitol (DDT); 10 g SDS; 10 mg bromophenol blue in a final volume of 100 ml; stored in 1-ml portions at −70 °C.
- 7.5% SDS–polyacrylamide running gel:

Solution	*Volume*	*Final concentration*
H_2O	20 ml	
1.5 M Tris-HCl, pH 8.5; 0.4% SDS	10 ml	0.375 M Tris, pH 8.5; 0.1% SDS
30% acrylamide, 0.8% bisacrylamide	10 ml	7.5% acrylamide, 0.2% bisacrylamide
10% ammonium persulphate	200 μl	0.05% ammonium persulphate

N,N,N′,N′-tetramethylethylenediamine (TEMED)	20 μl	0.005% TEMED

- 7.5% SDS–polyacrylamide stacking gel:

Solution	*Volume*	*Final concentration*
140 mM Tris-HCl, pH 6.8; 0.11% SDS	7 ml	0.125 M Tris, pH 6.8; 0.1% SDS
30% acrylamide, 0.8% bisacrylamide	1 ml	4.3% acrylamide; 0.1% bisacrylamide
10% ammonium persulphate	70 μl	0.25% ammonium persulphate
TEMED	7 μl	0.0025% TEMED

- fluorographic enhancement solution: 125 mM salicylic acid (sodium salt, pH 7.0); 30% methanol

Method

1. Add 10 μl 5 × GSB to each sample and mix by vortexing.
2. Heat samples 5 min at 95 °C.
3. Centrifuge briefly to bring down condensate.
4. Load on to 7.5% SDS–polyacrylamide gel and run at 25–50 mA, constant current.
5. Soak gel 20 min in fluorographic enhancement solution.
6. Dry gel.
7. Expose to Kodak XAR-5 film at −80 °C. Usually an overnight exposure is required.
8. Determine the fraction of G_D (*Figure 2*) by densitometry of the autoradiogram, using a GS300 transmission scanning densitometer (Hoefer Instruments) connected to an IBM-XT with the GS350 integrating software (Hoefer Instruments).

In a standard reaction, transport is generally 50–80% effective. The rate and extent of transport and cytosol dependence can vary between different preparations of semi-intact cells, so that the respective controls need to be done for every preparation.

3.4 *In vitro* transport to the endo D sensitive form in wild-type CHO cells

Transport to the endo D sensitive form can also be measured in wild-type semi-intact cells. The transport assay is exactly the same as described for 15B cells (see *Protocol 7*). A control sample has to be digested with endo H (*Protocol 10*), as a small fraction of VSV G-protein is further processed to an endo H resistant form in wild-type cells, even in the absence of externally added UDP–glcNAc. The percentage of endo H resistant VSV G-protein (which is again endo D resistant and cannot be distinguished from the endo D resistant ER-form) is added to the percentage of endo D sensitive form to calculate the fraction of G-protein that has been transported to the *cis* Golgi.

VSV G-protein that has been transported in wild-type CHO cells is digested with pure endo D. Endo D that is prepared from *D. pneumoniae* (22) may contain other glycosidases that also digest higher glycosylated oligosaccharides.

Protocol 10. Post-incubation with endo H

Material

- concentrated endo H buffer: 0.3% SDS; 0.1 M NaOAc, pH 5.6; 20 μl/ml β-mercaptoethanol (added just prior to use)
- 0.1 M NaOAc, pH 5.6
- endo H (Boehringer Mannheim, no. 1088 734)

Method

1. Pellet the cooled cells by centrifugation for 15 sec in a microcentrifuge at 15 000 *g*.
2. Resuspend the pellet in 20 μl concentrated endo H buffer.
3. Boil immediately (to prevent digestion of denatured G-protein by released lysosomal proteases) for 5 min.
4. Let the samples cool.
5. Add 40 μl 0.1 M NaOAc, pH 5.6 and 3 μl (3 mU) endo H.
6. Mix by vortexing.
7. Incubate overnight at 37 °C.

3.5 Interpretation of the data

The percentage of G-protein in each of the bands (G_{D0}, G_D) is determined by scanning of the films. *Figure 2* shows that at the beginning of the transport incubation, all of the G protein is found in the endo D resistant G_{D0} form. After a lag period of 5–15 min, the endo D sensitive G_D form appears, increases over a time period of about 60–80 min, and reaches a maximum after about 90 min of incubation.

Acknowledgements

The methods described here have been developed in the lab of William E. Balch. I would like to thank Steve Pind and Howard Davidson for helpful comments on the manuscript.

References

1. Gruenberg, J. E. and Howell, K. E. (1989). *Ann. Rev. Cell Biol.* **5**, 453.
2. Goda, Y. and Pfeffer, S. R. (1989). *FASEB J.* **3**, 2488.
3. Balch, W. E. (1989). *J. Biol. Chem.* **264**, 16965.
4. Rothman, J. E. and Orci, L. (1990). *FASEB J.* **4**, 1460.
5. Woodman, P. G. and Edwardson, J. M. (1986). *J. Cell Biol.* **103**, 1829.
6. Balch, W. E., Wagner, K. R., and Keller, D. S. (1987). *J. Cell Biol.* **104**, 749.
7. de Curtis, I. and Simons, K. (1989). *Cell* **58**, 719.
8. Tooze, S. A. and Huttner, W. B. (1990). *Cell* **60**, 837.
9. Fries, E. and Rothman, J. E. (1980). *Proc. nat. Acad. Sci., USA* **77**, 3870.
10. Haselbeck, A. and Schekman, R. (1986). *Proc. nat. Acad. Sci., USA* **83**, 2017.
11. Balch, W. E., Dunphy, W. G., Braell, W. A., and Rothman, J. E. (1984). *Cell* **39**, 405.
12. Rothman, J. E. (1987). *J. Biol. Chem.* **262**, 12502.
13. Beckers, C. J. M., Keller, D. S., and Balch, W. E. (1987). *Cell* **50**, 523.
14. Simons, K. and Virta, H. (1987). *EMBO J.* **6**, 2241.
15. Ruohola, H., Kabcenell, A. K., and Ferro-Novick, S. (1988). *J. Cell Biol.*, **107**, 1465.
16. de Curtis, L. and Simons, K. (1988). *Proc. nat. Acad. Sci., USA* **85**, 8052.
17. Baker, D., Hicke, L., Rexach, M., Schleyer, M., and Schekman, R. (1988). *Cell* **54**, 335.
18. Schwaninger, R., Beckers, C. J. M., and Balch, W. E. (1991). *J. Biol. Chem.* **266**, 13055.
19. Beckers, C. J. M. and Balch, W. E. (1989). *J. Cell Biol.* **108**, 1245.
20. Pind, S., Davidson, H., Schwaninger, R., Beckers, C. J. M., Plutner, H., Schmid, S. L., and Balch, W. E. (in press). In *Methods in enzymology*, Vol. 00. Academic Press.
21. Balch, W. E. and Rothman, J. E. (1985). *Arch. Biochem. Biophys.* **240**, 413.
22. Glasgow, L. R., Paulson, J. C., and Hill, R. L. (1977). *J. Biol. Chem.* **252**, 8615.

10

Lipid modifications involved in protein targeting

WAYNE J. MASTERSON and ANTHONY I. MAGEE

1. Introduction

Several contributions in this volume describe the role of protein sequence motifs in acting as determinants of targeting specificity. In the last 12 years or so a group of post-translational modifications of proteins with various lipid molecules has been identified in eukaryotic cells (1). In many cases these modifications have been shown to be essential for the subcellular localization of the proteins carrying them. Many intracellular proteins are now known to carry fatty acid modifications which serve such a purpose. The 14 carbon saturated fatty acid myristate (C14:0) is amide-linked to the amino-terminus of a range of proteins, exemplified by the $pp60^{src}$ (proto) oncogene product. Site-directed mutagenesis has been used to demonstrate that this acylation is required both for membrane localization and transforming activity (2). The myristoylation enzyme has been isolated and is highly specific for acyl chain length (3). Inhibitors which interfere with this pathway provide an attractive approach to specific chemotherapy of cancers and viral infections including HIV (4).

Longer chain fatty acids, predominantly palmitate (C16:0), can be linked to proteins, usually via thioester bonds to cysteine or oxyester bonds to serine or threonine. Palmitoylation can occur both on transmembrane proteins and on entirely intracellular proteins. In the former case, however the acylation sites are generally on the cytoplasmically disposed sequences or within the transmembrane region itself (5). Palmitoylation of *ras* proteins has been shown to be required for tight membrane binding and to co-operate with polyisoprenylation (see next paragraph) to determine specific subcellular localization (6, 7). The function of palmitoylation for transmembrane proteins is less obvious. However, palmitoylation is involved in the coupling of members of the β-adrenergic family to G proteins (8) and in intracellular routing of transferrin receptors (9). The dynamic turnover of palmitate on $p21^{ras}$ (10), transferrin receptor (11), and other proteins may be involved in regulating such aspects of protein function.

Polyisoprenylation of proteins by alkylation of cysteine with 15 carbon (farnesyl) or 20 carbon (geranylgeranyl) steroid precursors has been identified over the last few years. Farnesylation of *ras* proteins co-operates with nearby upstream palmitate residues or a polybasic sequence to specify plasma membrane binding (7). These modifications occur at carboxy-terminal motifs of the general type CAAX (C, cysteine; A, aliphatic; X, any other amino acid). Small and hydrophilic side chains in the X position favour fanesylation of the cysteine residue, while large aliphatic residues favour geranylgeranylation (12). In addition to polyisoprenylation these sequences are also modified by proteolytic removal of the AAX sequence, and by carboxyl-methylation of the α-carboxyl group (13, 14). Recent data suggest that carboxyl-terminal motifs of the type CysCys or CysXCys, found in the *ras*-related family of *rab* proteins, are modified in a related way (E. Fawell, J. F. Hancock, C. M. H. Newman, T. Giannakouros, J. Armstrong, and A. I. Magee, unpublished observations). The subtle variation in these lipid modifications, in concert with protein sequences, very likely contributes to the exquisite selectivity of different members of this family for compartments of the endomembrane system (15).

A wide variety of eukaryotic proteins are covalently attached to glycosyl phosphatidylinositol (GPI). The primary role of the GPI is to anchor the protein to the outer leaflet of the plasma membrane (16–18). In this regard, the GPI structure can be considered an alternative to the hydrophobic transmembrane polypeptide domain for protein anchorage. Detailed structures of four different GPI anchors have been published (19–22), from which a consensus GPI structure can be postulated (*Figure 1*). There is growing evidence that the GPI-membrane anchor can play a role in protein targeting. In polarized cells the presence of a GPI anchor is sufficient to direct a protein exclusively to the apical membrane (23, 24).

It is the purpose of this chapter to provide experimentally detailed protocols for studying the structure, biosynthesis, and function of each of these lipid modifications.

2. Protein acylation

In a few cases acylation has been recognized by structural analyses of unlabelled purified proteins (25). However, the modification of proteins with fatty acids is usually detected after metabolic labelling of appropriate cultured cell lines or primary cells with specific fatty acids. Due to the propensity of cells to interconvert fatty acids by chain elongation or shortening by *β*-oxidation the identity of the incorporated label must be confirmed. Distinction between amide- and thioester-linked fatty acids can be made by simple chemical cleavages.

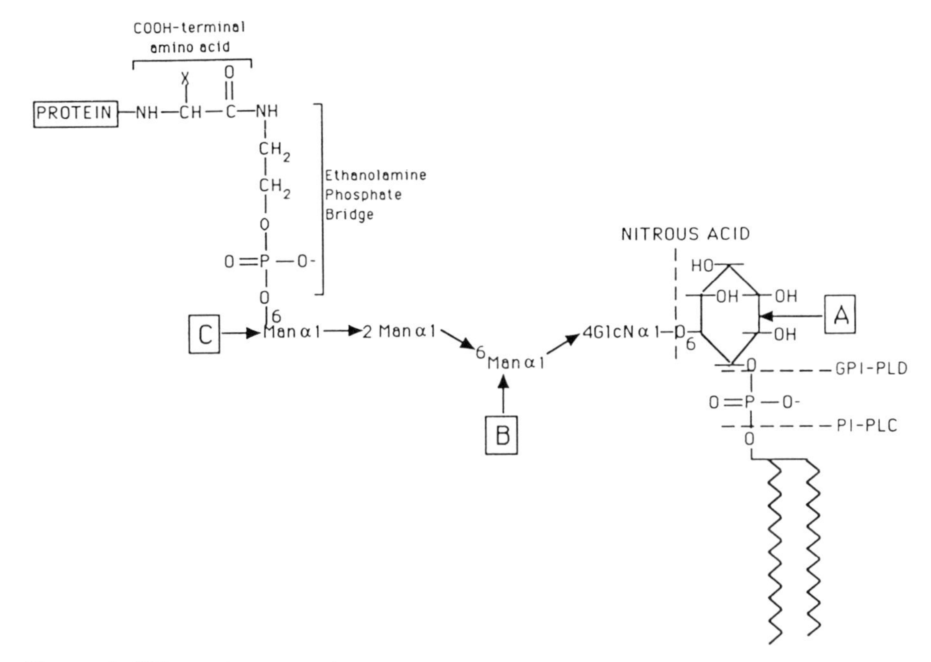

Figure 1. GPI-membrane anchor consensus structure. This structure is based on references 19–22. Boxed letters indicate various modifications of the GPI core structure. A is palmitate in ester-linkage to the inositol ring (21); B can be a side-chain of α-linked galactose residues (19), a ß-linked N-acetylgalactosamine residue (20), or phospho-ethanolamine (20, 21); C is an α-linked mannose residue (20). Dashed lines indicate sites of enzymatic or chemical cleavage.

2.1 Metabolic labelling with fatty acids

During the authors' early studies of protein acylation in cultured cells several variations of labelling medium were tried. These included the use of de-lipidated serum, pre-complexation of fatty acid with fatty acid-free bovine serum albumin (BSA), etc. Using extended labelling periods of several hours little effect of these treatments was observed. However, the use of a reduced serum concentration (1–5%) and the inclusion of 5 mM Na pyruvate and increased amino acid concentrations to reduce re-incorporation of label were found to be helpful. The authors's standard labelling medium (FA medium) for one to several hours is shown in *Table 1*.

The 9,10-tritiated fatty acids provide the best combination of high specific activity and detectability for *in vivo* labelling. Also, due to the tritium being far removed from the carboxyl end where ß-oxidation occurs, the reincorporation of label is minimized. For these reasons ^{14}C-fatty acids, especially those labelled in the 1-position, should be avoided for *in vivo* work. They can be used, however, for *in vitro* studies of acylation (26).

Table 1. FA medium for metabolic labelling of cultured cells

Dulbecco's modified Eagle's medium (DMEM) (or suitable alternative depending on cell type)
5% serum (donor, newborn, or fetal calf)
4 × the normal concentration of non-essential amino acids
5 mM Na pyruvate
50–200 μCi/ml [9, 10-^{3}H] palmitic acid (40–60 Ci/mmol, Amersham TRK 760) or [9, 10-^{3}H]myristic acid (10–30 Ci/mmol, New England Nuclear NET-830)

Fatty acids are often provided by the manufacturer as stocks in toluene. This should be removed under a gentle stream of nitrogen in a fume hood, and the fatty acids redissolved in ethanol at 10–50 μCi/μl. Under these conditions they can be stored for months or years at −20°C.

Relatively short labelling times (4–6 h) are usually chosen to reduce the problems of re-incorporation of label. This problem varies greatly with cell type; some cells show specific labelling even after 24 h. A useful control is to analyse total cell lysates by SDS–PAGE and compare the fatty acid-labelled samples with a [^{35}S]methionine-labelled sample. The labelling patterns with the fatty acids should be substantially different from each other and from the [^{35}S]methionine-labelled proteins (27). *Protocol 1* gives typical experimental conditions.

Protocol 1. Metabolic labelling of tissue culture cells with fatty acids

1. Grow cells in tissue culture dishes to 80–90% confluence. Alternatively, tissues can be rapidly removed from a newly-sacrificed animal, rinsed in serum-free tissue culture medium, and diced into small (~ 1 mm^2) pieces with fine scissors and a scalpel. The pieces are transferred into appropriate dishes in complete tissue culture medium.
2. Replace the medium with warm FA medium (*Table 1*) for 1 h.
3. Add [9, 10-^{3}H]fatty acids in ethanol directly to the medium (tilt the dish slightly and add the label to the 'deep end' to prevent the ethanol drying out the cells). Agitate gently to disperse the label.
4. Leave for the desired time (e.g. 6 h).
5. Wash the cells with cold phosphate-buffered saline (PBS) and lyse them directly in gel-loading buffer containing dithiothreitol (DTT) (28), or in immunoprecipitation buffer ('RIPA' 20 mM Tris; 150 mM NaCl; 1 mM EDTA; 0.5% (v/v) NP40; 0.5% (w/v) Na-deoxycholate; 0.1% (w/v) SDS; 1% (v/v) Aprotinin (Sigma); 0.2 mM phenylmethylsulphonyfluoride (PMSF), pH 7.4). The PMSF is kept as a 100 mM stock solution in isopropanol at room temperature and added just before use due to its instability in aqueous solution.

6. Load the lysates directly, or after immunoprecipitation with an appropriate antibody, on to SDS–polyacrylamide gels (28). Run the dye front off the gel to remove free label and phospholipids. Since some thioester-linked acyl groups are extremely labile, being removed even by reduction, the concentration of DTT in the gel loading buffer should not exceed 20 mM. The authors have found fluorography with 2, 5-diphenyloxazole (PPO) to be more efficient than with water-soluble fluors, e.g. salicylate. The authors use the PPO–DMSO method (29) or the PPO–acetic acid method (30) described in *Protocol 2*.

2.2 Pulse-chase labelling with fatty acids

Pulse-chase analysis is a valuable tool for the study of dynamic fatty acid acylation, especially for palmitate (10). Protocol 1 for long-term labelling of cells with fatty acids has to be modified for pulse-chase labelling. The pool size of fatty acids in cells is generally large compared to, for example, that of amino acids used in protein synthesis. This makes chasing of labelled fatty acids more difficult. Also, since pulse times are of necessity short, the presence of free fatty acids in the serum used in the medium will reduce the specific activity of the label and thus the counts incorporated. The authors have found it necessary to screen batches of serum for their ability to support efficient fatty acid labelling in short labelling periods, e.g. 5 min. A 'cloudy' appearance of the serum batch clearly indicates a high lipid content and should be avoided.

The incorporation of the labelled fatty acid can be optimized by previously starving the cells for up to 1 day in a medium containing a very low amount of serum. This allows the cells to use up their own endogenous fatty acid without affecting their metabolism markedly. In addition, the authors have observed that some cells are sensitive to repetitive changes of medium and we have had to adapt our method accordingly.

Protocol 2. Pulse-chase analysis of palmitoylation

1. Seed cells at a density of around 5×10^5 cells/cm^2 in medium containing 0.1% serum. After 1 day adjust the medium to the minimum volume (e.g. 1 ml for a 35-mm dish) and add cycloheximide (50 μg/ml; Sigma) for 1 h. This blocks protein synthesis and thus inhibits fatty acid incorporation into most acylated proteins but not into proteins whose acylation occurs post-translationally, e.g. p21^{ras}.
2. To each dish add 100 μCi/ml [9, 10-^{3}H]palmitate. First dilute the ethanol solution (10 μCi/μl) with medium at a ratio of 1:10 v/v, then add this to the dish containing 10 volumes of medium. This way of adding tritiated palmitate has proven to be efficient for metabolic incorporation without

Protocol 2. *Continued*

harming the cells. A time length of 15 min for the pulse was chosen for p21^{ras} following the authors observation of a time lag for incorporation of the label, corresponding probably to the time that palmitate needs to exchange between the medium and the cell.

3. After 15 min of incubation with label, gently remove the medium and replace with warm complete fresh medium (e.g. 2 ml containing 10% serum for the chase).
4. At each step of the chase cells from one dish are collected. Suck off the medium; rinse once with ice-cold PBS. Take the cells up into gel-loading buffer containing 10 mM DTT (e.g. 0.1 ml) or RIPA buffer (*Protocol 1*) for immunoprecipitation (e.g. 1 ml).
5. Resolve the resulting lysates of immunoprecipitates by SDS–PAGE and soak first in fresh glacial acetic acid for 5 min, then in 20% (w/v) PPO in glacial acetic acid for 90 min, and finally twice in distilled water for 20 min each time (30). This fluorographic treatment has proven to be more efficient and rapid than the traditional PPO–DMSO approach (29).
6. Incubate the gel in a solution of polyethylene glycol (PEG) 4000 (50% w/v) in water at 70 °C for 15 min (31). This induces a shrinkage of the gel by half and consequently increases the sensitivity of the fluorography.
7. Finally dry the gel at 60 °C for 2 h and fluorograph as usual.

2.3 Linkage analysis of fatty acid in acylated proteins

Distinction can be made between thioesters, oxyesters, and amides by chemical cleavage with selective agents. This is most conveniently performed on duplicate slices from an SDS–polyacrylamide gel of [^{3}H]fatty acid-labelled and [^{35}S]methionine-labelled protein(s) (*Protocol 3*). The method can be adapted for protein in solution. Linkage analysis should not be performed on fixed on dried gels since both these procedures can lead to transacylation converting labile ester linkages to stable linkages of unknown nature (possibly amides).

Protocol 3. Linkage analysis of fatty acylated proteins

1. Excise replicate lanes from a fresh gel and shake at room temperature for 1 h with 0.2 M KOH in methanol (this cleaves thio- and oxyesters but not amides; reference 32). Methanol alone is used to treat a control strip.
2. Treat a third strip with freshly prepared 1 M hydroxylamine hydrochloride titrated to pH 7.5 with NaOH (this rapidly cleaves thioesters, but cleaves oxyesters or amides only poorly; reference 33). Treat a fourth strip with

1 M Tris-HCl, pH 7.5, as a control. Hydroxylamine will cleave oxyesters at high pH (> 9).

3. Wash the slices three times with water for 5 min, and prepare for fluorography as in *Protocol 2*, steps **5–7**. Cleavage is measured as a reduction in the fluorographic signal compared to controls and can be quantitated by densitometric scanning or scintillation counting of excised bands.

Hydroxylamine treament in solution can also be used as a mild method for removal of thioester-linked fatty acid for functional studies (10) or to prepare deacylated acceptor protein for *in vitro* acylation studies (26).

Other diagnostic chemical cleavages include reduction of thioester to produce fatty alcohols (34) and thioester cleavage by methylamine (35).

2.4 Analysis of protein-bound label

Due to problems of interconversion of fatty acids via β-oxidation and chain elongation, and of reincorporation of label into other metabolic precursors, the protein-bound label derived from [^{3}H]fatty acids should be analysed (*Protocols 4* and *5*).

Protocol 4. Fatty acid analysis of acylated proteins in cell extracts

Total protein-bound label can be analysed using the protocol.

1. Precipitate labelled protein (e.g. cell lysate in 1% SDS) in a polypropylene tube with 5 volumes of acetone containing 0.1 M HCl at − 20 °C for at least 1 h, and pellet the precipitate by centrifugation.
2. Dissolve the pellet in a minimum volume of 1% SDS, transfer into a 1.5-ml microcentrifuge tube, and re-precipitate twice as in step **1**.
3. Extract the pellet at least three times by trituration with chloroform–methanol (2:1) until no more free label is extracted into the organic solvent as determined by scintillation counting.
4. Extract the pellet with diethyl ether, and dry under a nitrogen stream.
5. Place the microcentrifuge tube into a thick-walled Teflon container with a tight-fitting screw top (e.g. Tuftainer, Pierce) containing 1 ml 6 M HCl and flush with nitrogen. Close the lid tightly and hydrolyse in an oven at 110 °C for 16 h.
6. Extract the contents of the tube twice with 0.5 ml hexane, and pool the extracts. Determine the radioactivity in the hexane extracts and in the residue (after dissolving in 1% SDS). Fatty acids should be quantitatively extracted into hexane, while label reincorporated into sugars and amino acids will be mainly in the hexane residue.

Protocol 4. *Continued*

7. Evaporate the hexane extracts (e.g. in a Speed Vac), and dissolve in a small volume (2–5 μl) of chloroform–methanol (2:1). Spot on to a RP18 thin-layer plate (Merck 13724), and develop in acetonitrile–acetic acid (90:10). Spot authentic [^{3}H]fatty acids in parallel lanes as markers.
8. Dry the plate and detect the radioactivity by spraying with En^{3}hance spray (NEN) and exposure to preflashed Kodak XAR-5 film at −70 °C, or by scraping 1 cm lengths of adsorbent and scintillation counting.

Protocol 5. Fatty acid analysis of acylated proteins in gel bands

1. Following electrophoresis, locate the band of interest either by using molecular weight standards, or by fluorography using Na-salicylate (36). PPO-treated gels cannot be used.
2. Excise the band from the wet or dried gel, and wash three times with 0.5 ml water for 5 min each wash (agitate), during which time the dried gel piece rehydrates and the salicylate is washed out.
3. Transfer the gel piece to a 1.5-ml microcentrifuge tube, lyophilize, and proceed as for total protein (*Protocol 4*, step **5**). Alternatively, protein can be digested out of the gel with 200 μg/ml pronase (Calbiochem) in 20 mM ammonium bicarbonate, 0.05% SDS, pH 8.0, at 37 °C for 48 h, followed by lyophilization.
4. Hydrolyse the peptides, extract with hexane, and analyse by thin layer chromatography as in *Protocol 4*, steps **7** and **8**.

---

3. Protein polyisoprenylation and carboxyl-methylation

As discussed in Section 1, the covalent attachment of polyisoprenoid moieties to proteins has recently been observed and is associated with intracellular targeting of those proteins. Frequently, carboxyl-terminal proteolytic processing and carboxyl-methylation accompany polyisoprenylation.

Polyisoprenoids such as farnesyl (15 carbon) and geranylgeranyl (20 carbon) are derived from mevalonic acid via a pathway which results ultimately in the production of a wide range of cellular metabolites, e.g. steroids, dolichols, ubiquinone, haem (37). The intermediates are produced as activated pyrophosphate derivatives which are polymerized starting from the 5 carbon precursor isopentenyl pyrophosphate. The first step in the pathway is the production of mevalonic acid from hydroxymethylglutaryl–coenzyme A (HMG–CoA) catalysed by the enzyme HMG–CoA reductase. Several highly specific inhibitors of this enzyme are available and are used clinically as anti-hypercholesterolaemic agents. Some of these are:

- Mevinolin (lovastatin): Merck, Sharp, and Dohme
- Compactin (ML-236B): Sankyo Co.
- Pravastatin: E. R. Squibb and Sons Limited

Inhibition of HMG–CoA reductase by these agents causes depletion of cytosolic mevalonic acid, resulting in the accumulation of non-isoprenylated precursor proteins. The lack of isoprenylation frequently causes a decrease in hydrophobicity which can be assessed by phase partitioning using the detergent Triton X-114. [^{3}H]mevalonic acid added to treated cells is incorporated into the accumulated precursors, which can be detected as radiolabelled bands on SDS–PAGE.

3.1 Preparation and use of mevinolin

The authors have used mevinolin (38) which is supplied in the lactone form and can be hydrolysed to the free acid (*Protocol 6*).

Protocol 6. Preparation of mevinolin (39)

1. Dissolve mevinolin (90 mg) in ethanol (1.8 ml) at 55 °C.
2. Add 0.9 ml of 0.6 M NaOH and 18 ml distilled water.
3. Incubate at room temperature for 30 min.
4. Adjust to pH 8 with HCl and dilute to the desired concentration. A stock solution can be stored indefinitely at −20 °C as aliquots and the authors have found no deleterious effects of freezing and thawing, at least for up to six times.

For cultured cells the authors have found medium concentrations of 25–50 μM mevinolin to be highly effective within 1 h of treatment. Cells are transferred to FA medium (Table 1) and the mevinolin added directly to the medium from the concentrated stock.

3.2 Triton X-114 partitioning

This method, pioneered by Bordier (40), takes advantage of the ability of aqueous solutions of the non-ionic detergent Triton X-114 (TX-114) to separate into two phases at 25–30 °C, due to its low cloud point. Hydrophilic, soluble proteins tend to partition into the upper aqueous phase, while hydrophobic, membrane proteins tend to partition into the lower detergent phase (*Protocol 8*). Before use hydrophilic contaminants of the commercially available TX-114 must be removed by pre-condensation (*Protocol 7*).

Protocol 7. Pre-condensation of Triton X-114

1. Dissolve the TX-114 (Sigma) at 30 g/l in buffer (e.g. Tris-buffered saline) on ice.
2. Add a trace of bromophenol blue to stain the detergent phase.
3. Warm the solution to 37 °C and centrifuge.
4. Discard the upper phase, and redissolve the detergent-enriched phase in buffer on ice.
5. Repeat steps **3** and **4** a further three times.
6. Store the detergent-enriched phase as a stock at 4 °C (for longer storage freeze at −20 °C). When condensed at 37 °C in the presence of 150 mM NaCl, TX-114 forms a phase containing approximately 12% detergent.

Protocol 8. Triton X-114 phase partitioning

Materials

A 1% stock solution of TX-114 is made up in an appropriate buffer (e.g. 50 mM Tris-HCl, 150 mM NaCl, 1 mM EDTA, pH 7.5, containing 0.2 mM PMSF and 1% (v/v) Aprotinin (Sigma))

Method

1. Grow cells as a monolayer and metabolically label as desired, e.g. with [^{35}S]methionine or Tran[^{35}S] label (ICN). Mevinolin treatment (25–50 μM, ≥ 1 h pre-treatment) can be performed if desired.
2. Wash the monolayer twice in ice-cold PBS, lyse in 1% cold TX-114 (1 ml for a 35-mm Petri dish), scrape with the blunt end of a yellow pipetman tip, and transfer to a 1.5-ml microcentrifuge tube. Keep on ice for 10 min.
3. Pellet the insoluble debris in a refrigerated microcentrifuge for 2 min. Transfer the supernatant to a fresh tube.
4. Place the tube in a water bath at 30 °C for 2 min, during which time the detergent phase separates.
5. Centrifuge for 1 min in a microcentrifuge at room temperature.
6. Remove the upper aqueous phase (~ 0.9 ml) to a second tube. Add 0.9 ml of buffer to the lower detergent phase and 0.1 ml of 12% TX-114 stock to the aqueous phase and place on ice until a homogeneous solution is obtained (~ 5 min). (At this stage the phase separation can be repeated to clean the phases further. However, if care is taken during the removal of the aqueous phase the authors do not usually find this to be necessary.)

7. Prepare the separated phases for analysis by immunoprecipitation in the usual way, or by trichloroacetic acid (TCA) precipitation. For this, add TCA from a 100% (w/v) stock in water to a final concentration of 10% (w/v) for ≥ 1 h at 4 °C (10 μg of ovalbumin can be added to the detergent phase sample to assist in precipitation). Collect the pellet by microcentrifuging, wash twice with ice-cold ethanol, once with diethyl ether, and air-dry. Dissolve the pellet in gel loading buffer. If the bromophenol blue indicator in the loading buffer turns yellow as an indication of residual acid, 1-μl aliquots of 1 M NaOH should be added until it just becomes blue. The sample can then be resolved by SDS–PAGE and analysed by Western blotting.

3.3 Mevalonic acid labelling

R-[5-^{3}H]mevalonic acid (New England Nuclear NET-716; ~ 35 Ci/mmol) is supplied in 50% aqueous ethanol. Just before use the ethanol should be removed by a gentle stream of nitrogen or in a SpeedVac. The mevalonic acid must not be heated above 20 °C since it is unstable. Having removed the ethanol, the aqueous solution is added directly to cells pre-incubated in FA medium containing 25–50 μM mevinolin for ≥ 1 h, at a concentration of 50–200 μCi/ml for up to 16 h. At the end of the incubation the cells are processed for analysis by immunoprecipitation and SDS–PAGE.

For highest sensitivity the authors use PPO as a fluor. However, if further analysis of the resolved radioactive bands is to be performed (see *Protocol 9*), fluorography is done with Na salicylate (36). Typical exposure times are 1 week to 1 month.

3.4 Isoprenoid analysis of incorporated radiolabel

Having observed incorporation of label from [^{3}H]mevalonic acid into a protein it is important to determine the chain length of the polyisoprenoid moiety. This is because at least two types of polyisoprenoid have been found to modify proteins, namely farnesyl and geranylgeranyl, and the nature of this substituent can affect the properties of the protein as well as point towards the pathway of modification.

TCA precipitates of proteins can be used, but great care must be taken to remove non-covalently bound polyisoprenoids. The TCA precipitate (obtained as in *Protocol 8*) is delipidated by trituration thrice with acetone in the cold (e.g. on a vortex mixer in a cold room) for at least 6 h each time. Proteins resolved by SDS–PAGE have already been effectively delipidated and can be analysed by the method of Casey *et al* (41) (*Protocol 9*). Polyisoprenoids can be conveniently analysed by HPLC after cleavage from the protein by methyl iodide.

Protocol 9. Polyisoprenoid cleavage

1. Excise salicylate-fluorographed bands from a dried gel, using the fluorogram as a guide. In order to have enough counts for analysis, a strong fluorographic signal should be obtained in 1 week. If longer exposures are required then multiple gel pieces should be used.
2. Place the gel piece in a 1.5-ml microcentrifuge tube and wash thrice for 5 min each time with double distilled water at room temperature to remove salicylate and SDS. Remove the paper backing from the gel at the same time.
3. Add 100 µl 0.1 M ammonium bicarbonate containing 0.02% Na azide and 10 µg pronase (Calbiochem) to the gel piece. Incubate overnight at 37 °C. This digests the protein in the gel and aids elution.
4. Remove the supernatant and redigest with a further 100 µl NH_4HCO_3 and 10 µg pronase for 6 h at 37 °C. Remove this supernatant and re-extract for 2 h at 37 °C with 200 µl buffer alone.
5. Pool the three extracts in a 10-ml polypropylene tube and lyophilize. Dissolve the residue in 0.5 ml water and lyophilize again.
6. Dissolve the residue in 400 µl 25 mM Tris-HCl, pH 7.7 in 80% HPLC grade acetonitrile (degassed by nitrogen flushing to remove oxygen), and leave 1 h at room temperature. Add 800 µl N_2-degassed 3% (w/v) formic acid followed by 100 µl fresh methyl iodide. Flush with nitrogen and keep in the dark overnight at room temperature with vortex mixing.
7. Remove the methyl iodide under gentle vacuum (e.g. with a water pump), then add 150 µl 35% (w/v) Na_2CO_3, flush with nitrogen and leave overnight in the dark with gentle mixing.
8. Extract the cleaved polyisoprenoids twice with 1.2 ml Analar chloroform–methanol (9:1). Pool the extracts and count the organic extract and the aqueous residue.

The authors have typically achieved a total recover of ~ 50% of the counts with 30–40% of the recovered counts in the organic phase. This is improved somewhat by nitrogen degassing as described above. Ideally control samples should be processed identically, but with the omission of methyl iodide.

3.4.1 HPLC analysis of polyisoprenoids

Following cleavage, dry the chloroform-methanol pool under a stream of nitrogen and redissolve in 100 µl of HPLC-grade acetonitrile containing 25 mM phosphoric acid (solvent B). It is important to use a high acetonitrile concentration since geranylgeraniol is poorly soluble in aqueous solvents. Just before HPLC analysis dilute the sample with an equal volume of HPLC-grade water, vortex vigorously, and microcentrifuge to remove particulate material.

The authors have used a Beckman System Gold HPLC and a Brownlee Aquapore RP300 cartridge (4.6 × 100 mm) coupled with a 3-cm guard cartridge of the same material. Any octadecyl (or shorter carbon chain) reverse phase column could be used for this purpose with appropriate optimization of the separation.

Inject the sample (usually half) on to the column equilibrated in 50% aqueous acetonitrile containing 25 mM phosphoric acid (solvent A) at a flow rate of 0.5 ml/min. Solvent B is 100% acetonitrile–25 mM phosphoric acid. The gradient used for elution is:

- 10 min with 100% solvent A
- linear gradient to 100% solvent B over 10 min
- 10 min hold at 100% solvent B
- 5 min reverse gradient to 100% solvent A

Collect 1-min fractions in scintillation vial inserts (the authors use a Pharmacia FRAC-100 collector) and count after the addition of 5 ml scintillation fluid.

Standard isoprenoids can be obtained from the following sources:

- geraniol—Sigma
- farnesol—Sigma or Aldrich
- geranylgeraniol—Kuraray Co. Ltd (contact Dr M. Tanomura)

Keep stock solutions (1 mM) of each isoprenoid in acetonitrile at −20°C. Check the elution positions daily by injecting a mixture of the three (1 nmol each) in solvent A and monitoring the elution at 254 nm. *Figure 2* shows a typical elution pattern and retention times.

Staring with 10–20 000 c.p.m. of gel eluted material the authors have obtained peaks in the HPLC analysis of ~ 200 c.p.m. using half the sample. Given the improvements achieved by N_2 degassing of solvents, one can estimate that a minimum of 1000–2000 c.p.m. of starting material would be required for an analysis, injecting all the sample.

3.5 Analysis of protein carboxyl-methylation

The final stage in processing CAAX boxes, after polyisoprenylation and proteolytic trimming of AAX, is carboxyl-methylation of the α-carboxyl group at the C-terminus. This is conveniently assayed *in vivo* by the incorporation of label from [methyl-^{3}H]methionine into alkali-labile ester bonds. Some of the methionine is metabolically converted into S-adenosyl methionine (SAM), the donor for methylation reactions in the cell. The labelled methyl group can then be incorporated into both alkali-stable methyl groups (e.g. of lysine, arginine, and histidine) or alkali-labile methyl esters.

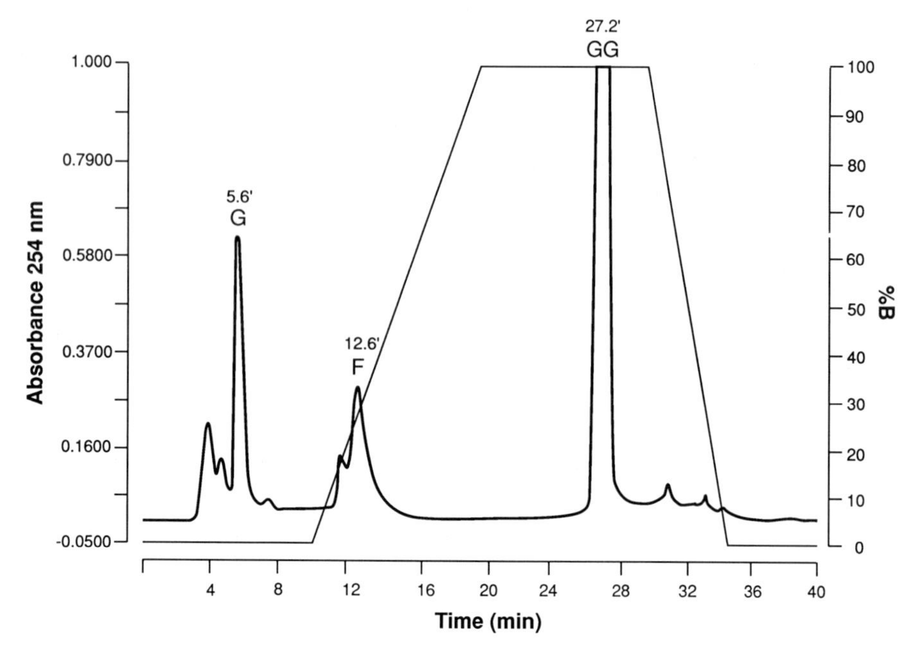

Figure 2. HPLC analysis of polyisoprenoid standards (see for details). G, geraniol; F, farnesol; GG, geranylgeraniol. Retention times are shown.

Labelled methionine will also be incorporated directly into the primary structure of proteins during protein synthesis. C-terminal carboxyl-methyl groups cannot easily be distinguished from methyl esters formed with the side chains of aspartate and glutamate residues, although these are somewhat more labile. However, if the presence of a CAAX sequence is known then it strongly indicates a carboxyl-terminal methylation.

The authors' method for analysis of carboxyl-methylated proteins is based on that of Chelsky *et al.* (42) (*Protocol 10*).

Protocol 10. Carboxyl-methyl analysis

1. Grow cells to subconfluent density on a Petri dish (e.g. 35 mm) and incubate overnight with low methionine medium (5% the normal concentration of methionine) containing 50–200 μCi/ml L-[methyl-^{3}H] methionine (Amersham TRK 584; ~ 80 Ci/mmol). Methionine-free medium can be used for shorter labelling times.
2. Prepare lysates and run directly on SDS–PAGE or immunoprecipitate as usual. The pH of the buffers should be kept below 7.5 to avoid demethylation.

3. After SDS–PAGE treat the gel with salicylate and fluorograph (36).
4. Excise the labelled bands using the fluorogram as a guide and place in the bottom of 1.5-ml microcentrifuge tubes.
5. Place the open microcentrifuge tubes upright inside 20-ml glass scintillation vials containing 7 ml scintillation fluid (see *Figure 3*).
6. Add 100 μl 1 M NaOH to the tube and rapidly cap the scintillation vial. Place overnight at 37 °C. During this time [^{3}H]methanol is hydrolysed from the protein and distils into the surrounding scintillation fluid.
7. The next day carefully remove the microcentrifuge tube (taking care not to spill any contents into the scintillation vial). Add an equal amount of 1 M HCl to the contents of the tubes and transfer to a fresh scintillation vial. Add 7 ml scintillation fluid and count the samples. The volatile methanol counts measure carboxyl-methylation, while the non-volatile counts remaining in the tube measure the incorporation of methionine directly into protein.

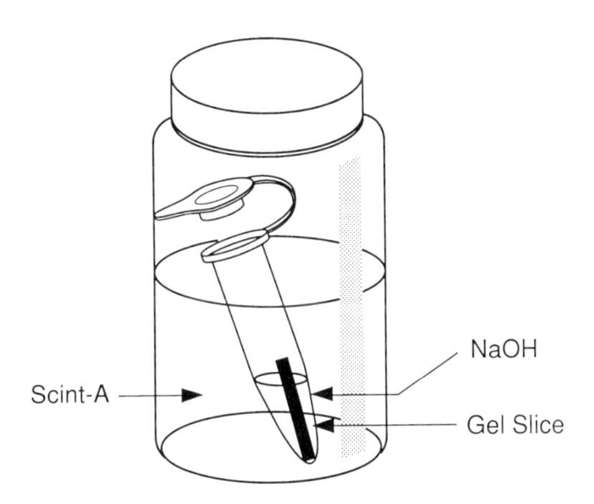

Figure 3. Carboxyl-methyl assay (see text for details).

Assuming that the specific activity of the SAM pool and of the methionyl-tRNA pool are similar and that methylation is tightly coupled to translation, then the ratio of methanol counts to methionine counts is an estimate of the extent of methylation. Also, if the primary sequence and therefore the number of methionines in the protein is known, the stoichiometry of methylation can be calculated. This value is usually close to one (e.g. for p21$^{N\text{-}ras}$, reference 13).

4. Glycosylphosphatidylinositol (GPI) anchors

The best way of identifying a GPI-membrane anchor is by direct chemical analysis (see reference 43), but this requires at least 10 nmol of purified starting material. There are, however, a number of techniques that allow the identification of low-abundance GPI-membrane anchored proteins, starting with whole cells or membrane preparations. Since no single technique is unequivocal, a number are outlined here and their limitations discussed.

4.1 cDNA sequence analysis

cDNA sequence provides the first indication of a GPI-membrane anchor. GPI-membrane anchors (16–18) are attached to protein in the endoplasmic reticulum (ER). The addition of GPI requires the presence of (a) a signal sequence, to direct the nascent polypeptide across the ER membrane, and (b) a C-terminal GPI addition signal peptide (GPIsp) which is rapidly cleaved and directly replaced by a pre-assembled GPI precursor. There is no defined consensus sequence for GPIsp, rather a set of general features (*Figure 4*).

(a) The known residues of the GPIsp cleavage/GPI attachment site are serine, aspartate, asparagine, cysteine, and glycine. Hydrophobic and aromatic amino acids are excluded. The +1 and +2 position residues of the cleaved GPIsp so far include combinations of alanine, asparagine, aspartate, glycine, and serine. The regions found N-terminal of the cleavage site show no obvious homologies.
(b) Following the cleavage site is a moderately polar domain of about 7–12 residues. These regions contain no significant homologies.
(c) The GPIsp generally terminates in a row of 10–20 random hydrophobic amino acids (16–18, 44, 45).

Figure 4. Features of the GPI-addition signal peptide. Arrow indicates the site of cleavage (and hence the COOH-terminus of the mature proteins). Numbers are referred to in the text.

(d) The clearest indication of a GPIsp is the absence of a typical cytoplasmic domain (i.e. sequences containing at least two positively charged residues) immediately following the hydrophobic domain. There is one exception to this rule, that of Qa-2, where the C-terminus appears like a transmembrane domain followed by a tripeptide containing two arginine residues. However, the transmembrane-like domain is interrupted in the middle by a negatively charged aspartate residue and conversion of the residue to valine abolishes GPIsp activity (46). Short polar domains (containing single negatively (47) or positive (48) charged residues) immediately following the hydrophobic domain can be tolerated but these examples are rare or artificial (47).

4.2 Identification of the GPI-membrane anchor by PI–PLC cleavage

GPI-membrane anchors can be cleaved by bacterial phosphatidylinositol-specific phospholipase C (PI–PLC). This cleavage can be detected by a number of methods.

(a) *Release of antigen from the cell surface.* The alkyl and acyl chains of the GPI anchor the protein to the membrane. Removal of these chains by PI–PLC (see *Figure 1*) releases the protein from the cell surface into the surrounding medium. This release can be detected in a number of ways (see point (b)). A limitation of this technique is that some proteins with PI–PLC susceptible GPI-anchors are not released well from cell surfaces by this enzyme (e.g. 5′-nucleotidase) (17). The reasons for this are unclear but probably include steric hindrance of the enzyme (perhaps through association of the GPI-anchored protein with other proteins). This limitation can be circumvented by working with detergent extracts (see point (b)).

(b) *Change in detergent partitioning properties.* GPI-anchored membrane proteins bind non-ionic detergents because of their amphiphilic nature and should therefore partition with the detergent during phase partition using TX-114 (Section 3.2, *Protocol 8*). After the removal of the lipid moiety by PI–PLC cleavage the GPI protein should behave as an hydrophilic molecule and partition into the detergent-depleted aqueous phase.

(c) *Exposure of GPI-specific epitope (the cross-reacting determinant).* A problem with both Methods (a) and (b) is that they only provide indirect proof of a GPI-membrane anchor. A protein released by PI–PLC from the cell surface or into the aqueous phase of the phase partition could be *associated* with a GPI-membrane anchored protein, rather than having a GPI of its own. Fortunately, PI–PLC cleavage of the GPI-membrane anchor results not only in the release of alkyl/acyl glycerol (or diacylglycerol) but also in the formation of a 1,2-cyclic phosphate ring on the inositol residue. The inositol 1,2-cyclic phosphate, along with the non-N-acetylated glucosamine, forms the major part of an epitope known as the cross-reacting determinant (CRD; *Figure 5*) (49). As outlined in *Protocol 13* (and see *Figure 6*), following SDS–PAGE and transfer to nitrocellulose, GPI-membrane anchored proteins can be detected by probing with anti-CRD antibodies (following PI–PLC treatment on the nitrocellulose; *Protocol 12*). As a control for the specificity of antibody binding, blots can be given a mild acid treatment prior to the anti-CRD incubation. The mild acid treatment decyclises the 1,2-cyclic phosphate thus destroying the CRD epitope. An advantage of this method is that it allows you to use samples prepared in SDS–PAGE sample buffer. Such samples avoid a potential problem, the loss of the GPI-membrane anchor through proteolysis. The major disadvantage of this approach is that, since the CRD appears to be non-immunogenic in mice and rats, monoclonal antibodies against this epitope are not available. This has necessitated the use of polyclonal antisera in all CRD immunological studies to date.

4.2.1 Conditions for cell-surface release

Preparations of cells are incubated with *Bacillus cereus* (Boehringer 1143 069) or *Bacillus thuringiensis* PI-PLC to release GPI-anchored proteins. These enzymes can be used in the presence of most conventional buffers and tissue culture media at pH 7.0–7.4. Although the sensitivity of different substrates varies considerably, maximal release of most GPI-anchored proteins is given by 1 unit/ml (1 unit of enzyme hydrolyses 1 μmol PI/min) for 30–60 min at 37 °C.

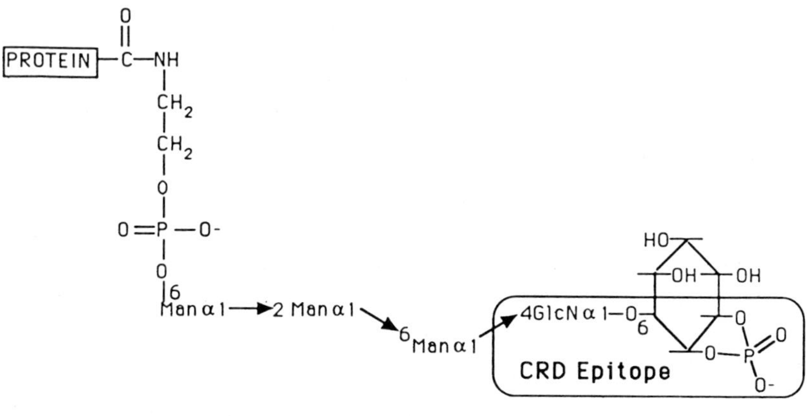

Figure 5. The cross-reacting determinant (CRD). The figure shows a simple anchor (22) after PI-PLC cleavage. The CRD epitope is boxed.

4.2.2 Analysis of protein release

(a) If an appropriate antibody is available, protein release can be readily assessed by comparing immunofluorescent staining of PI–PLC treated cells with controls. If the antigen is GPI-membrane anchored, PI–PLC treatment should diminish the staining intensity (usually measured by FACS analysis). Alternatively, released protein can be immunoprecipitated or detected by Western blotting.

(b) A range of hydrolases are GPI-membrane anchored. In these cases PI-PLC treatment releases enzyme activity into the supernatant. Examples include alkaline phosphatase, 5′-nucleotidase, renal dipeptidase, and acetylcholinesterase.

(c) Approaches (a) and (b) need a suitable probe (an antibody or an enzyme activity). A conceptually simple non-selective approach, which should be applicable to any tissue or cell type has been developed by Rodriguez-Boulan and colleagues (50). The surface of the cell is labelled using sulpho-*N*-hydroxysuccinamide-biotin. Since this reagent is membrane-impermeant, only proteins on the cell surface are labelled with biotin (*Protocol 11*). GPI-membrane anchored proteins released from the cell-surface by PI–PLC can be detected with ^{125}I-labelled streptavidin after SDS–PAGE and transfer to nitrocellulose (see *Protocol 12*).

Protocol 11. Cell-surface labelling with biotin

1. Wash cells five times in ice-cold phosphate-buffered saline containing 1 mM $CaCl_2$ and 1 mM $MgCl_2$.
2. Treat cells in the same buffer with 0.5 mg/ml sulpho-*N*-hydroxysuccinamide-biotin (Pierce) for 30 min at 4 °C.
3. Wash four times with ice-cold buffer.
4. Incubate cells with PI–PLC as described in Section 4.2.1.

Protocol 12. TX-114 extraction of cells and PI–PLC digestion

The example given in this protocol is for cells surface-labelled with biotin (see *Protocol 11*)

1. Extract cells (0.5–1.0 × 10^7) in 1 ml Tris-buffered saline (TBS) containing 1% TX-114 for 1 h at 4 °C with occasional vortexing; clarify and phase separate (see *Protocol 8*). Disard the upper aqueous phase.
2. Repartition the detergent phase twice.
3. Dilute the re-extracted detergent phase to a final volume of 0.5 ml with 0.1 M Tris-HCl; 50 mM NaCl; 1 mM EDTA, pH 7.4. Divide this equally between two tubes. To one tube add PI–PLC to a final concentration of

Protocol 12. *Continued*

6 units/ml (the other tube acts as a control). Allow the PI–PLC treatment to proceed for 1 h at 37 °C with continuous mixing.

4. Following PI–PLC treatment, dilute the samples to a final volume of 0.5 ml with TBS containing 2% TX-114 and phase-separate. Retain the aqueous phase (which contains the PI–PLC cleaved GPI-proteins).
5. Re-extract the aqueous phase three times by each time adding 0.05 ml 10% TX-114 in TBS (without EDTA) followed by phase separation.
6. The final aqueous phase from step **5** may still contain amphiphilic proteins which will confuse the analysis. To minimize these contaminants, after step **5** add Phenyl–Sepharose (0.2 ml of a 75% slurry per ml of sample; Pharmacia) and rotate (20 h, 4 °C).
7. Remove the Phenyl–Sepharose by centrifugation. Acetone or TCA precipitate the final supernatant (see *Protocol 8*).
8. Dissolve the precipitated protein in SDS–PAGE sample buffer, separate by SDS–PAGE, and transfer to nitrocellulose.
9. Incubate nitrocellulose sheets in blocking buffer (PBS containing 0.5% (v/v) Tween 20, 1 M D-glucose, 10% (v/v) glycerol, and 0.3% (w/v) bovine serum albumin (BSA)) for 2 h at 37 °C. Remove the blocking buffer and incubate with ^{125}I-labelled streptavidin (Amersham; 1–2 × 10^6 c.p.m./ml) in the same buffer for 2 h at 37 °C. Wash three times for 5 min at 37 °C.
10. Biotinylated proteins are detected by autoradiography on Kodak XAR-5 film (4–24 h at −80 °C with an intensifying screen).

Note. Protease inhibitors should be included at all stages to avoid loss of the GPI-anchor through proteolysis.

4.2.3 Detection of CRD-positive antigens following PI–PLC digestion of nitrocellulose

The principles of this approach are outlined in *Figure 5*. The conditions given (*Protocol 13*) are those used for detecting GPI-anchored proteins in the protozoal parasite *Trypanosoma cruzi* (L. Güther, personal communication), but the same procedure should be applicable to any cell type.

Protocol 13. Detection of CRD-positive antigens

1. Suspend 1.5 × 10^8 cells of the metacyclic stage of *T. cruzi* directly in 100 µl SDS–PAGE sample buffer (containing 0.1 M DTT or 5% (v/v) 2-mercaptoethanol) and boil for 3 min.

2. Analyse the sample (loaded in five tracks) by SDS–PAGE (10% resolving gel) and transfer to nitrocellulose paper by standard procedures.
3. After transfer, stain one of the tracks (along with a molecular weight standard track) for protein (e.g. with amido black). Wash the nitrocellulose bearing the other four tracks three times for 10 min each in blocking buffer (PBS, 3% (w/v) BSA, 0.25% (v/v) Tween 20).
4. Rinse the nitrocellulose twice in enzyme buffer (50 mM Tris-acetate, pH 7.2; 0.16% (w/v) sodium deoxycholate; 3% (w/v) BSA), and divide in two. Incubate one-half with shaking in the same buffer containing *B. thuringiensis* PI–PLC at 2.5 units/ml for 2 h at 37 °C (or overnight at 23 °C); the other half is incubated in buffer alone.
5. Rinse the two pieces of nitrocellulose with water and divide in two (now four pieces in all). Treat one from each pair with 1 M HCl (or 1 M NaCl as a mock) for 90 min at 23 °C. After rinsing well with water, wash the nitrocellulose (two times for 10 min each) with blocking buffer, the incubate for 1 h at 37 °C with anti-CRD (serum diluted 1:500 in blocking buffer). The antigen the authors use for anti-CRD antibody production is the PI–PLC cleaved form of *T. brucei* variant surface glycoprotein (VSG). Anti-CRD antibodies are commercially available from Oxford Glycosystems.
6. After washing three times for 10 min each with blocking buffer, bound antibody is detected by a number of standard procedures (e.g. goat anti-rabbit IgG-peroxidase conjugate of ^{125}I-labelled protein A). As shown in *Figure 6* a GPI-membrane anchored protein is one which binds anti-CRD antibodies only after PI–PLC treatment and this binding is abolished by mild acid treatment.

4.2.4 Limitations of the PI–PLC approach

Interpretation of results where only a proportion of a particular antigen appears to be PI–PLC sensitive is a problem. A good example is human erythrocyte acetylcholinesterase (E^{hu}AChE), only 10–15% of which is cleaved by PI–PLC (21, 51). The PI–PLC resistance could have been caused by an alternative mode of anchoring (e.g. an hydrophobic transmembranc peptide), but detailed analysis revealed that, in fact, all E^{hu}AChE has a GPI-membrane anchor. PI–PLC resistance results from acylation of the GPI inositol moiety (modification A in *Figure 1*). In the case of E^{hu}AChE, the small proportion susceptible to PI–PLC allowed the GPI to be detected. However, a protein with a GPI-membrane anchor which was completely of the acylated inositol type would not be detected by any of the methods so far described. This problem can be overcome if the GPI-membrane anchor can be biosynthetically radiolabelled (see Section 4.3). Once a radiolabel has

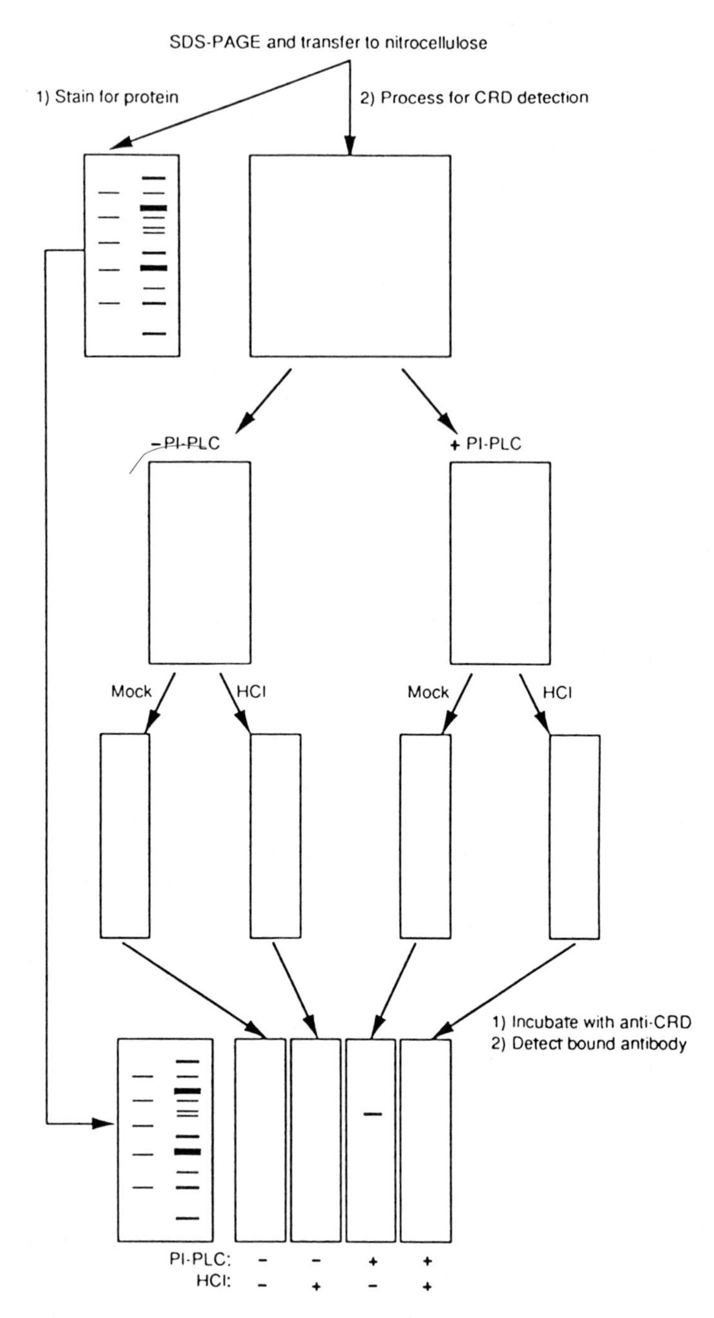

Figure 6. Strategy for identifying GPI-membrane anchored proteins using anti-CRD antibodies (see text for details).

been incorporated, a more detailed chemical and enzymatic analysis of the putative GPI structure is possible.

4.3 Metabolic labelling with GPI anchor components

The second approach to identifying a GPI-membrane anchor is by metabolic labelling with radiolabelled precursors of anchor components. To do this requires cells that are biosynthetically active. I will discuss just two cases (labelling with either [^{3}H]ethanolamine or [^{3}H]fatty acid) but GPI-membrane anchors have also been biosynthetically labelled with [^{3}H]inositol, [^{3}H]glucosamine, and [^{3}H]mannose (16–18).

4.3.1 Biosynthetic labelling with [^{3}H]ethanolamine

Ethanolamine is perhaps the best component of the anchor to use in biosynthetic studies. No special media are needed and, with only one known exception, the only way [^{3}H]ethanolamine can be incorporated into protein is through a GPI-membrane anchor. That one exception is protein synthesis elongation factor 1α (52–4), which is easily identified as an hydrophilic protein of about 50 kda.

The conditions outlined (*Protocol 14*) are those used in reference 54. These conditions have been used for a variety of cell types. As a control for labelling use a cell line known to express a GPI-membrane anchored protein (e.g. murine lymphocyte line expressing Thy-1).

Protocol 14. Metabolic labelling with ethanolamine

1. Suspend the cells in medium 199 with Earle's salts (Gibco) supplemented with penicillin (100 units/ml), streptomycin (0.1 mg/ml), 1 mM sodium pyruvate, 2 mM glutamine, BME amino acids (Gibco), and 10% fetal calf serum (FCS) (previously dialysed against PBS and filter-sterilized).
2. Add [^{3}H]ethan-1-ol-2-amine hydrochloride (5–30 Ci/mmol; Amersham TRK 462) at a concentration of 0.1 mCi/ml and culture in a humidified incubator at 37 °C in 5% CO_2 for 12–18 h.
3. Pellet the cells by centrifugation and wash with PBS.
4. Lyse the cells by suspending directly in SDS–PAGE sample buffer and boil for 3 min.
5. Analyse samples by SDS–PAGE and visualize labelled proteins by fluorography using salicylate (36).

Note: It is essential to verify (steps **6** to **8**) that the label biosynthetically incorporated into protein is indeed [^{3}H]ethanolamine.

6. Following SDS–PAGE, determine the position of the labelled protein with respect to pre-stained weight standards and excise that region of the gel.

Protocol 14. *Continued*

7. Transfer the gel piece to a microcentrifuge tube and hydrolyse (*Protocol 4*).
8. Analyse the hydrolysate on an amino acid analyser, or dansylate and analyse by TLC (55).

4.3.2 Biosynthetic labelling with [^{3}H]fatty acid

[^{3}H]Fatty acid biosynthetically incorporated into a GPI-anchor (see *Protocol 1*) can be readily distinguished from other lipid modifications of protein by a combination of enzymatic and chemical cleavages. This characterization must be done on material that is free of interfering [^{3}H]fatty acid-labelled lipids. SDS–PAGE, followed by electroelution of the band of interest, is probably the best way of isolating the [^{3}H]fatty acid-labelled protein free of contaminating lipid (55).

Protocol 15. Characterization of fatty-acid labelled GPI

A. *PI–PLC treatment*

Cleavage of [^{3}H] fatty-acid labelled protein by PI–PLC releases [^{3}H] diacyl (or alkylacyl) glycerol which is identified by TLC.

1. Add 0.2 ml 20 mM Tris-acetate, pH 7.5 and 0.1% Triton X-100 to each of two tubes containing 10 000 c.p.m. [^{3}H] protein. Add 0.2 units PI–PLC to one of the tubes. Incubate both tubes at 37 °C for 1 h.
2. Add 0.5 ml water-saturated *n*-butanol to both tubes; vortex and centrifuge to separate the phases.
3. Remove the upper phase and determine the % radioactivity released with respect tô the no-enzyme control by scintillation counting of an aliquot. Dry the remainder of the upper phase and redissolve in standard mix A (10 mg/ml each in chloroform of monomyristin, 1,2-dimyristin, 1,3-dimyristin, myristic acid, trimyristin, methylmyristate). Analyse by TLC.

B. *GPI-specific phospholipase D (GPI-PLD)*

GPI-membrane anchors which are resistant to PI–PLC cleavage because of an acylated inositol ring are susceptible to cleavage by GPI-PLD (and also nitrous acid—see part C). Serum is the source of this enzyme (56, 57). The labelled product of GPI-PLD treatment is [^{3}H] phosphatidic acid.

1. Add 0.2 ml 50 mM Tris-HCl; 10 mM NaCl; 2.6 mM $CaCl_2$; 0.1% Triton X-100, pH 7.4 to each of two tubes containing 10 000 c.p.m. [^{3}H] protein. Add 1 μl non-heat-inactivated rabbit serum to one of the two tubes. Incubate both tubes at 37 °C for 4 h.

2. Perform step **2** of part A.
3. Remove the upper phase and extract the lower phase with a further 0.5 ml water-saturated *n*-butanol. Combine the upper phases.
4. Determine the % radioactivity as in step **3** of part A. Dry the residue and redissolve in 20 μl standard mix B (10 mg/ml each in chloroform:methanol (2:1) of phosphatidic acid, phosphatidylcholine, phosphatidylethanolamine, phosphatidylglycerol, phosphatidylinositol). Analyse by TLC.

C. *Nitrous acid deamination*

Nitrous acid deamination of the non-N-acetylated glucosamine in the GPI-membrane anchor results in release of PI. In the case of PI–PLC resistant GPI-membrane anchors, the PI obtained has the extra acyl group of the inositol ring giving it a faster mobility than normal PI on TLC.

1. Suspend two 10 000-c.p.m. aliquots of [^{3}H] protein in 75 μl 0.1 M sodium acetate buffer, pH 4.0. To one, add 75 μl freshly prepared 0.5 M $NaNO_2$ and to the other 75 μl 0.5 M NaCl. Incubate at room temperature for 3 h.
2. Extract twice with 0.5 ml water-saturated *n*-butanol. Wash the combined upper phases with 0.5 ml *n*-butanol-saturated water.
3. Perform step **4** of part B.

D. *TLC analysis*

1. Activate Silica gel G or Si 60 TLC plates at 125 °C for 1 h before use. Equilibrate filter-paper-lined TLC tanks with solvent[a] for about 1 h before use.
2. After development, the standards are detected by placing the chromatograms in a tank of iodine vapour. ^{3}H-labelled products are detected with a TLC linear analyser or by scraping 0.5-cm strips into scintillation fluid.

[a] The solvent used for PI–PLC products is petroleum ether:diethyl ether:acetic acid (80:20:1). The solvent used for GPI-PLD and nitrous acid products is chloroform:methanol: methanol:acetic acid:water (25:15:4:2).

References

1. Magee,. A. I. (1990). *J. Cell Sci.* **97**, 581.
2. Kamps, M. P., Buss, J. E., and Sefton, B. M. (1986). *Cell* **45**, 105.
3. Towler, D. A., Adams, S. P., Eubanks, S. R., Towery, D. S., Jackson-Machelski, E., Glaser, L., and Gordon, J. I. (1987). *Proc. nat. Acad. Sci., USA* **84**, 2708.
4. Bryant, M. L., Heuckeroth, R. O., Kimata, J. T., Ratner, L., and Gordon, J. I. (1989). *Proc. nat. Acad. Sci., USA* **86**, 8655.

5. Schmidt, M. F. G. (1989). *Biochim. Biophys. Acta* **988**, 411.
6. Hancock, J. F., Magee, A. I., Childs, J. E., and Marshall, C. J. (1989). *Cell* **57**, 1167.
7. Hancock, J. F., Paterson, H., and Marshall, C. J. (1990). *Cell* **63**, 133.
8. O'Dowd, B. F., Hnatowich, M., Caron, M. G., Lefkowitz, R. J., and Bouvier, M. (1989). *J. Biol. Chem.* **264**, 7564.
9. Alvarez, E., Gironès, N., and Davis, R. J. (1990). *J. Biol. Chem.* **265**, 16644.
10. Magee, A. I., Gutierrez, L., McKay, I. A., Marshall, C. J., and Hall, A. (1987). *EMBO J.* **6**, 3353.
11. Omary, M. B. and Trowbridge, I. S. (1981). *J. Biol. Chem.* **256**, 12888.
12. Yamane, H. K., Farnsworth, C. C., Xie, H., Evans, T., Howald, W. N., Gelb, M. H., Glomset, J. A., Clarke, S., and Fung, B. K.-K. (1991). *Proc. nat. Acad. Sci., USA.* **88**, 286.
13. Gutierrez, L., Magee, A. I., Marshall, C. J., and Hancock, J. F. (1989). *EMBO J.* **8**, 1093.
14. Clarke, S., Vogel, J. P., Deschenes, R. J., and Stock, J. (1988). *Proc. nat. Acad. Sci., USA.* **85**, 4643.
15. Chavrier, P., Parton, R. G., Hauri, H. P., Simons, K., and Zerial, M. (1990). *Cell* **62**, 317.
16. Ferguson, M. A. J. and Williams, A. F. (1988). *Ann. Rev. Biochem.* **57**, 285.
17. Low, M. G. (1989). *Biochim. Biophys. Acta* **988**, 427.
18. Cross, G. A. M. (1990). *Ann. Rec. Cell Biol.* **6**, 1.
19. Ferguson,M. A. J., Homans, S. W., Dwek, R. A., and Rademacher, T. W. (1988). *Science* **239**, 753.
20. Homans, S. W., Ferguson, M. A. J., Dwek, R. A., and Rademacher, T. W. (1988). *Nature* **333**, 269.
21. Roberts, W. L., Santikarn, S., Reinhold, V. N., and Rosenberry, T. L. (1988). *J. Biol. Chem.* **263**, 18776.
22. Schneider, P., Ferguson, M. A. J., McConville, M. J., Mehlert, A., Homans, S. W., and Bordier, C. (1990). *J. Biol. Chem.* **265**, 16955.
23. Lisanti, M. P., Caras, I. W., Davitz, M. A., and Rodriguez-Boulan, E. (1989). *J. Cell Biol.* **109**, 2145.
24. Dotti, C. G., Parton, R. G., and Simons, K. (1991). *Nature* **349**, 158.
25. Aitken, A., Cohen, P., Santikarn, S., Williams, D. H., Calder, A. G., Smith, A., and Klee, C. B. (1982). *FEBS Lett.* **150**, 314.
26. Berger, M. and Schmidt, M. F. G. (1984). *J. Biol. Chem.* **259**, 7245.
27. Magee, A. I. and Courtneidge, S. A. (1985). *EMBO J.* **4**, 1137.
28. Laemmli, U. K. (1970). *Nature* **227**, 680.
29. Laskey, R. A. (1980). In *Methods in enzymology*, Vol. 65 (ed. L. Grossman and K. Maldave), pp. 363–71. Academic Press, New York.
30. Skinner, M. K. and Griswold, M. D. (1983). *Biochem. J.* **209**, 281.
31. Mohamed, M. A., Lerro, K. A., and Prestwick, G. D. (1989). *Anal. Biochem.* **177**, 287.
32. Schmidt, M. F. G., Bracha, M., and Schlesinger, M. J. (1979). *Proc. nat. Acad. Sci., USA* **76**, 1687.
33. Magee, A. I., Koyama, A. H., Malfer, C., Won, D., and Schlesinger, M. J. (1984). *Biochim. Biophys. Acta* **798**, 156.
34. Ross, N. W. and Braun, P. E. (1988). *J. neurosci. Res.* **21**, 35.

35. Bizzozero, O. A. and Good, L. K. (1990). *J. Neurochem.* **55**, 1986.
36. Chamberlain, J. P. (1979). *Anal. Biochem.* **98**, 132.
37. Maltese, W. A. (1990). *FASEB J.* **4**, 3319.
38. Alberts, A. W., Chen, J., Kuron, G., Hunt, V., Huff, J., Hoffman, C., Rothrock, J., Lopez, M., Joshua, H., Harris, E., Patchett, A., Monaghan, R., Currie, S., Stapley, E., Albers-Schonberg, G., Hensens, O., Hirshfield, J., Hoogsten, K., Liesch, J., and Springer, J. (1980). *Proc. nat. Acad. Sci., USA* **77**, 3957.
39. Kita, T., Brown, M. S., and Goldstein, J. L. (1980). *J. clin. Invest.* **66**, 1094.
40. Bordier, C. (1981). *J. Biol. Chem.* **256**, 1604.
41. Casey, P. J., Solski, P. A., Der, C. J., and Buss, J. E. (1989). *Proc. nat. Acad. Sci., USA* **86**, 8323.
42. Chelsky, D., Gutterson, N. I., and Koshland, D. E. (1984). *Anal. Biochem.* **141**, 143.
43. Ferguson, M. A. J. (1992). In *Lipid modification of proteins: a practical approach* (ed. N. M. Hooper and A. J. Turner), pp. 191–230. IRL Press, Oxford.
44. Berger, J., Howard, A. D., Brink, L., Gerber, L., Hauber, J., Cullen, B. R., and Udenfriend, S. (1988). *J. Biol. Chem.* **263**, 10016.
45. Caras, I. W., Weddell, G. N., and Williams, S. R. (1989). *J. Cell Biol.* **108**, 1387.
46. Waneck, G. L., Stein, M. E., and Flavell, R. A. (1988). *Science* **241**, 697.
47. Caras, I. W. and Weddell, G. N. (1989). *Science* **243**, 1196.
48. Scallon, B. J., Scigliano, E., Freedman, V. H., Miedel, M. C., Pan, Y.-C. E., Unkeless, J. C., and Kochan, J. P. (1989). *Proc. nat. Acad. Sci., USA* **86**, 5079.
49. Zamze, S. E., Ferguson, M. A. J., Collins, R., Dwek, R. A., and Rademacher, T. W. (1988). *Eur. J. Biochem.* **176**, 527.
50. Lisanti, M. R., Sangiacomo, M., Graeve, L., Saltiel, A. R., and Rodriguez-Boulan, E. (1988). *Proc. nat. Acad. Sci., USA* **85**, 9557.
51. Roberts, W. L., Myher, J. J., Kuksis, A., Low, M. G., and Rosenberry, T. L. (1988). *J. Biol. Chem.* **263**, 18766.
52. Tisdale, E. J. and Tartakoff, A. M. (1988). *J. Biol. Chem,* **263**, 8244.
53. Rosenberry, T. L. *et al.* (1989). *J. Biol. Chem.* **264**, 7096.
54. Whiteheart, S. W., Shenbagamurthi, P., Chen, L., Cotter, R. J., and Hart, G. W. (1989). *J. Biol. Chem.* **264**, 14334.
55. Haldar, K., Ferguson, M. A. J., and Cross, G. A. M. (1985). *J. Biol. Chem.* **260**, 4969.
56. Davitz, M. A., Hereld, D., Shak, S., Krakow, J., Englund, P. T., and Nussenzweig, V. (1987). *Science* **238**, 81.
57. Low, M. G. and Prasad, A. R. (1988). *Proc. nat. Acad. Sci., USA* **85**, 980.

A1

Suppliers of specialist items

Items referred to in the text can be purchased from the following companies. In general, the address of the head office is listed, from which the names and addresses of more local suppliers can be obtained.

Agar Aids, 66A Cambridge Road, Stanstead, Essex CM24 8DA, UK.

Aldrich Chemical Co., 940 West Saint Paul Avenue, Milwaukee, WI 53233, USA.

American Type Culture Collection, 12301 Parklawn Drive, Rockville, MD 20852, USA.

Amersham International plc, Amersham Place, Little Chalfont, Amersham, Bucks HP7 9NA, UK.

Associated Biomedic Systems Inc., 872 Main Street, Buffalo, NY 14202, USA.

BDH Laboratory Supplies, Broom Road, Poole, Dorset BH12 4NN, UK.

Beckman Instruments Inc., 2500 Harbor Boulevard, Box 3100, Fullerton, CA 92634, USA.

Becton Dickinson Labware, PO Box 7375, Mountain View, CA 94039, USA.

Bellingham and Stanley Ltd, Polyfract Words, Longfield Rd, Tunbridge Wells, Kent TN2 3EY, UK.

Bernitech Engineering, P.O. Box 3592, Saratoga, CA, 95070 USA.

Bio-Rad Laboratories, 3300 Regatta Boulevard, Richmond, CA 94804, USA.

Boehringer Mannheim GmbH Biochemica, Sandhofer Str. 116, Postfach 310120, D-6800 Mannheim 31, Germany.

Branson, Ultrasonics Corp., 41 Eagle Rd., Danbury, CT 06813–1961, USA.

Brownlee, Labs. Inc., 2045 Martin Ave., #204, Santa Clara, CA 95050, USA.

Calbiochem. Corp., PO Box 12087, San Diego, CA 92112–4180, USA.

Cambridge Bioscience, 25 Signet Court, Newmarket Road, Cambridge CB5 8LA, UK.

Cetus, Perkin-Elmer Corp, 761 Main Ave, Norwalk, CT 06859–0012, USA.

Costar, Cambridge, MA, USA.

Coulter Electronics Inc., PO Box 2145, MC195–10, Hialeah, FL33012–0145, USA.

De Fonbrune Metallurgical Service Laboratories Ltd, Reliant Works, Brockham, Surrey RH3 7HW, UK.

Dow Corning Corp., Medical Products, Midlands, MI 48640, USA.

DuPont Company Biotechnology Systems, Barley Mill Plaza, Chandler Mill Building, Wilmington, DE 19898, USA.

Dynal (UK) Ltd, Station House, 26 Grove Street, New Ferry, Wirral, Merseyside L62 5AZ, UK.

Eastman-Kodak Co., 343 State Street, Rochester, NY 14650, USA.

Eppendorf Inc., 45636 Northport Loop East, Fremont, CA 94538, USA.

Fisher Scientific, 711 Forbes Ave, Pittsburgh, PA 15219, USA.

Fisons Scientific Equipment, Bishops Meadow Road, Loughborough, Leicestershire LE11 0RG, UK.

Fungal Genetic Stock Center, Donner Laboratory, University of California, Berkeley, CA 94720, USA.

Gibco-BRL, Life Technologies Limited, PO Box 35, Trident House, Renfrew Road, Paisley PA3 4EF, Scotland, UK.

Hoefer Scientific Instruments, 654 Minnesota St, San Franscisco, CA 94107–9985, USA.

Arnold R. Horwell Ltd, 73 Maygrove Road, West Hampstead, London NW6 2BP, UK.

IBF Biotechnics, 35 Avenue J. Jaurés, 92390 Villeneuve-la-Garonne, Paris, France.

ICN Biomedicals, Free Press House, Castle St, High Wycombe, Bucks HP13 6RN, UK.

Intervet Laboratories, Intervet UK Limited, Science Park, Milton Road, Cambridge CB4 4FP, UK.

Janssen Pharmaceutical Ltd, Grove, Oxfordshire OX12 0DQ, UK.

Kuraray Co. Ltd, Chemicals Division II, 12–39 Umeda, 1-chome, Kita-ku, Osaka 530, Japan.

LKB Produkter AB, Box 305, S-16126 Bromma, Sweden.

E. Merck, Franfurter Strasse 250, D-6100 Darmstadt 1, Germany.

Merck, Sharp, and Dohme Research Labs, Merck & Co. Inc., PO Box 2000, Rahway, NJ 07065, USA.

Millipore Corporation, 80 Ashby Road, Bedford, MA 02118, USA.

Molecular Probes, Inc, 4849 Pitchford Avenue, Eugene, OR 97402, USA.

New England Nuclear, 549 Albany Street, Boston, MA 02118, USA.

Novo Enzyme Products, 9 Lion and Lamb Yard, West St, Farnham, Surrey GU9 7U, UK.

Nycomed (UK) Ltd, Nycomed House, 2111 Coventry Road, Sheldon, Birmingham B26 3EA, UK.

Oxford Glycosystems, Unit 4, Hitching Court, Blacklands Way, Abingdon, Oxon OX14 1RG, UK.

Parr Instruments Co., 211 53rd St. Moline, IL 61265, USA.

Pharmacia-LKB Biotechnology AB, S75182 Uppsala, Sweden.

Pierce Biochemicals, PO Box 117, Rockford, IL 61105, USA.

Portex Ltd, Hythe, Kent CT21 6JL, UK.

Promega Corp., 2800 South Fish Hatchery Road, Madison, WI 53711–5305, USA.

Research Instruments Ltd, Kernick Road, Penryn, Cornwall TR10 9DQ, UK.

Riedel de Haen, D-1035, Seelze 1, Germany.

Röhm Pharma GmbH, Postfach 4347, D6100, Darmstadt 1, Germany.

Sankyo Co. Ltd, Tokyo 140, Japan.

Sarstedt Ltd, 68 Boston Rd, Beaumont Leys, Leicester LE4 1AW, UK.

Schleicher and Schüll GmbH, Postfach 4, D-3354 Dassel, Germany.

Scientific and Medical Products Ltd, Shirley Institute, Didsbury, Manchester M20 8RX, UK.

Seikagaku America Inc., 9049 4th Street North, PO Box 21517, St Petersburg, FL 33742, USA.

Serva Feinbiochemica GmbH, PO Box 105260, Carl-Benz. Str. 7, D-6900 Heidelberg 1, Germany.

Sigma Chemical Co., PO Box 14508, St Louis, MO 63178, USA.

Singer Instruments Co. Ltd, Treborough Lodge, Roadwater, Watchet, Somerset TA23 0QL, UK.

E. R. Squibb and Sons Ltd, The Squibb Institute for Medical Research, Research Chemical Distribution Centre, PO Box 4000, Princeton, NJ 08540, USA.

Stratagene Cloning Systems, 11099 Torrey Pines Road, La Jolla, CA 92037, USA.

Taab Laboratories Equipment Ltd, 3 Minerva House, Callera Industrial Park, Aldermaston, Berks RG7 4QW, UK.

Thomas Scientific, 99 Hill Rd, 1/295 Box 99, Swedesboro, NJ 080581/0099, USA.

Vickers Medical Instruments, Priestley Road, Basingstoke, Hants RG24 9NP, UK.

Watson-Marlow Ltd, Falmouth, Cornwall, TR11 4RU, UK.

Whatman Ltd, Springfield Mill, Maidstone, Kent ME14 2LE, UK.

Index